Cities and Nature

Series Editors

Peter Newman, Sustainability Policy Institute, Curtin University, Perth, Australia

Cheryl Desha, School of Engineering and Built Environment, Griffith University, Nathan, Australia

Alessandro Sanches-Pereira, Instituto 17, São Paulo, Brazil

Cities and Nature fosters high-quality multi-disciplinary research addressing the interface between cities and the natural environment. It provides a valuable source of relevant knowledge for researchers, planners and policy-makers. The series welcomes empirically based, cutting-edge and theoretical research in urban geography, urban planning, environmental planning, urban ecology, regional science and economics. It publishes peer-reviewed edited and authored volumes on topics dealing with the urban and the environment nexus, including: spatial dynamics of urban built areas, urban and peri-urban agriculture, urban greening and green infrastructure, environmental planning, urban forests, urban ecology, regional dynamics and landscape fragmentation.

Indexed in Scopus!

Bijeesh Kozhikkodan Veettil · Bhumika Das
Editors

Remote Sensing of Urban Heat Islands

Case Studies from Emerging Cities

Springer

Editors
Bijeesh Kozhikkodan Veettil
Laboratory of Ecology and Environmental Management, Science and Technology Advanced Institute (STAI)
Van Lang University
Ho Chi Minh City, Vietnam

Bhumika Das
Department of Civil Engineering
Lakhmi Chand Institute of Technology
Bilaspur, Chhattisgarh, India

ISSN 2520-8306 ISSN 2520-8314 (electronic)
Cities and Nature
ISBN 978-3-032-11579-9 ISBN 978-3-032-11580-5 (eBook)
https://doi.org/10.1007/978-3-032-11580-5

This Springer imprint is published by the registered company Springer Nature Switzerland AG
The registered company address is: Gewerbestrasse 11, 6330 Cham, Switzerland

Preface

This book aims to provide a broad overview of urban heat island research from a remote sensing perspective. Most of the chapters are written by scientists researching urban heat islands (UHI) in India, Vietnam, and Brazil. Chapter "Understanding Urban Heat Islands: Definition, Causes and Effects" overviews urban heat islands, their definition, causes, and impacts. A historical perspective of urban heat island research and the significance of UHI research are discussed in Chapter "Historical Perspective: Urban Heat Islands (UHIs) Research and Its Significance". Various technologies and instruments, including satellites and sensors used for UHI research, are discussed in Chapter "Technologies and Instruments: Satellites and Sensors for Urban Heat Island Research". Chapter "Thermal Remote Sensing: A Case Study on Detecting and Analysing Surface Temperature Variations in Response to Land Use and Land Cover Changes" describes thermal remote sensing in UHI studies, including a case study from Brazil. Chapter "Land Use Land Cover Analysis: Understanding the Relationship Between Urban Development and Urban Heat Islands" demonstrated the relationship between urban development and the formation and expansion of UHI. A detailed picture of the impacts of LULC variations on land surface temperature variations is provided with the help of a case study in Chapter "Assessing the Impact of LULC on Land Surface Temperature Variations Using GIS and Remote Sensing". A scientometrics analysis of the global trends in UHI research with the help of remote sensing is given in Chapter "Global Trends in Urban Heat Island Research Using Remote Sensing: A Scientometrics Analysis". A case study on vegetation degradation, urban sprawl, and land surface temperatures from Vietnam is given in Chapter "Vegetation Degradation, Urban Sprawl, and Land Surface Temperature Variations in the Dalat City, Vietnam". The role of artificial intelligence (AI) in urban heat island research is discussed in Chapter "Role of Artificial Intelligence (AI) in Urban Heat Island Research". Chapter "Policy Implications of Integrating Remote Sensing Data into Urban Policy and Planning" provides the policy implications of integrating remote sensing data into urban policy and planning.

Ho Chi Minh City, Vietnam — Bijeesh Kozhikkodan Veettil
Bilaspur, India — Bhumika Das

Contents

Understanding Urban Heat Islands: Definition, Causes, and Effects

V. Anitha

Abstract Urban Heat Islands (UHIs) represent a significant environmental challenge, marked by elevated temperatures in urban areas compared to their rural counterparts. This phenomenon arises primarily due to urbanization, land use changes, and intensified human activities. UHIs have wide-ranging implications, including increased energy demand, adverse effects on public health, and degradation of air and water quality, as well as impacts on urban ecosystems. This chapter provides a comprehensive analysis of UHIs, focusing on both Surface Urban Heat Islands (SUHIs) and Atmospheric Urban Heat Islands (AUHIs). It delves into their definition, underlying causes, and multifaceted effects, emphasizing their role in shaping urban sustainability. Understanding and mitigating the UHI effect is essential for creating climate-resilient cities and fostering long-term urban sustainability.

Keywords Urban heat islands · Urban sprawl · Land surface temperature · Urban planning · Sustainable development

1 Introduction

The rapid expansion of urban areas worldwide has introduced significant environmental challenges, with the Urban Heat Island (UHI) effect emerging as a critical concern. First documented in the nineteenth century, the UHI effect describes the phenomenon where urban regions consistently record higher temperatures than their rural surroundings, particularly during nighttime. This temperature disparity is driven by factors such as changes in land use, intensive energy consumption, and the unique design of urban environments. Understanding the UHI effect is essential as cities grow, given its implications for public health, energy demand, and environmental sustainability (Urban Heat Islands 101—Resources for the Future).

V. Anitha (✉)
Former Ph.D. Research Scholar, Department of Environmental Sciences, Bharathiar University, Coimbatore, Tamil Nadu, India
e-mail: anithavijayakumar8@gmail.com

B. Kozhikkodan Veettil and B. Das (eds.), *Remote Sensing of Urban Heat Islands*, Cities and Nature, https://doi.org/10.1007/978-3-032-11580-5_1

A comprehensive exploration of UHIs includes an examination of their two primary forms: Surface Urban Heat Islands (SUHIs), which are linked to heat retention in urban materials, and Atmospheric Urban Heat Islands (AUHIs), characterized by elevated air temperatures in urban regions. The underlying causes of UHIs, including reduced vegetation, anthropogenic heat release, and the thermal properties of urban structures, contribute to their formation. These factors lead to far-reaching effects, such as increased energy consumption, elevated health risks, and disruptions to urban ecosystems (U.S. EPA 2008).

This discussion addresses the definition, types, causes, and effects of the UHI phenomenon, laying the groundwork for understanding a pressing environmental issue in the context of urbanization and climate change.

2 Definition of Urban Heat Islands

Urban Heat Islands (UHIs) refer to the phenomenon in which urban areas experience higher temperatures than their surrounding rural regions. This temperature disparity is most pronounced during the night and is influenced by a combination of natural and anthropogenic factors. The intensity of UHIs can vary based on the city's size, geographic location, and weather conditions. This phenomenon is measured by assessing surface and atmospheric temperature differences, with satellite imagery and ground-based observations frequently used for analysis (Oke 1982).

The temperature difference is primarily caused by urbanization and human activities that transform the natural landscape. Permeable surfaces and vegetation are replaced with impervious materials like asphalt and concrete, which absorb and retain heat. Reduced vegetation, the thermal properties of urban materials, urban geometry, and anthropogenic heat emissions all contribute to this effect. UHIs have significant implications for energy consumption, air quality, water systems, and human health, especially during heatwaves, exacerbating both environmental and societal challenges.

Surface Urban Heat Islands (SUHIs): Surface Urban Heat Islands (SUHIs) result from the absorption and retention of solar energy by urban surfaces such as pavements and rooftops. These surfaces can heat up to 50–90°F (27–50 °C) above the ambient air temperature on sunny days. SUHIs are typically strongest during the day and slightly weaker at night, with more pronounced effects in the summer due to higher solar intensity. Scientists often detect SUHIs through remote sensing techniques and thermal imaging. This type of UHI affects local air temperatures and contributes to atmospheric UHIs, intensifying urban temperature disparities (US EPA—www.epa.gov),

Atmospheric Urban Heat Islands (AUHIs): Atmospheric Urban Heat Islands (AUHIs) occur when air temperatures in urban areas exceed those in rural surroundings. These heat islands have two types: canopy layer UHIs (near the ground and below rooftops) and boundary layer UHIs (extending above the rooftops). AUHIs are most pronounced at night due to the slow release of heat stored in urban infrastructure, with temperature differences typically ranging from 1.8 to 5.4°F (1–3 °C). Detection relies on ground-based methods such as weather stations and mobile temperature measurements. AUHIs affect urban energy demands, health, and climate resilience (U.S. EPA 2017—www.epa.gov).

Comparison of SUHIs and AUHIs: While SUHIs are related to surface temperatures and are strongest during the day, AUHIs involve air temperature differences and peak at night. SUHIs have larger temperature variations and are detected using remote sensing, whereas AUHIs show more moderate differences and are measured directly. Despite their differences, both types contribute to urban warming and require mitigation strategies to reduce their impacts.

3 Causes of Urban Heat Islands

1. **Altered Land Use and Land Cover**: Urbanization replaces natural landscapes such as forests and grasslands with impervious surfaces like asphalt, concrete, and buildings. These materials absorb and retain heat more efficiently than natural surfaces, contributing to elevated urban temperatures (Grimmond 2007).
2. **Reduction in Vegetation**: Vegetation plays a crucial role in cooling through shading and evapotranspiration. Urban areas often experience significant vegetation loss, reducing the natural cooling effect and exacerbating heat retention (Gill et al. 2007).
3. **Heat Emissions from Human Activities**: Transportation, industrial processes, and energy consumption release heat into the environment. These anthropogenic heat sources intensify the UHI effect, particularly in densely populated and industrialized areas (Santamouris 2015).
4. **Urban Geometry and Design**: Tall buildings and narrow streets, common in urban environments, trap heat by limiting air circulation and creating heat-retaining urban canyons. This phenomenon, known as the "urban canyon effect," amplifies the UHI intensity (Rizwan et al. 2008).
5. **Reduced Albedo**: Urban surfaces often have low albedo, meaning they absorb rather than reflect sunlight. Dark-colored materials like asphalt contribute significantly to heat absorption, further intensifying UHIs (Akbari et al. 2001).
6. **Waste Heat from Air Conditioning**: Air conditioning systems discharge waste heat into the surrounding environment, creating localized hotspots and contributing to the overall urban heat balance (Salamanca et al. 2014).

4 Effects of Urban Heat Islands

1. **Increased Energy Consumption**: Higher temperatures drive increased demand for air conditioning, leading to higher energy consumption and associated greenhouse gas emissions. This creates a feedback loop that further exacerbates the

UHI effect (Santamouris 2014). A review of case studies from various countries revealed that electricity demand for air conditioning rises by about 1–9% with each 2°F increase in temperature. The highest increase was seen in countries like the United States, where air conditioning is widespread (Santamouris 2020). Parker and Barkaszi (1997) have demonstrated that applying high albedo coating to a building's roof results in a reduction of air conditioning energy use by 2–43% in houses.

2. **Impact on Urban Ecosystems**: UHIs significantly alter urban ecosystems by modifying microclimates, impacting biodiversity, and disrupting natural cycles. The ecological consequences of the UHI phenomenon include increased outbreaks of invasive species (Mogi et al. 2020), extended growing seasons (Zipper et al. 2016), shifts in insect phenology (Villalobos-Jimenez and Hassall 2017), and changes in the morphology of ectotherms (Magura et al. 2020).
3. **Decreased Air and Water Quality**: Increased temperatures exacerbate air pollution by accelerating the formation of ground-level ozone and smog. Similarly, runoff from heated impervious surfaces can degrade water quality in urban water bodies (U.S. EPA 2008).

   Increased energy consumption results in increased emissions of air pollutants and GHGs, which include nitrogen oxide, mercury, sulfur dioxide, and black carbon, which are all examples of pollutants (U.S. EPA 2024).
4. **Economic Impacts**: The financial burden of UHIs is evident in increased energy costs, healthcare expenses, and infrastructure maintenance. Urban heat also negatively impacts worker productivity and exacerbates socioeconomic disparities (Taha 1997).
5. **Public Health Impacts**: Elevated urban temperatures significantly increase the risk of heat-related illnesses like heatstroke and cardiovascular issues. Vulnerable populations, including the elderly and children, are particularly at risk during heat waves (McMichael et al. 2006). Rising temperatures and poor air quality in UHIs pose serious health consequences. The most critical health impact of the UHI effect is heat-related mortality (Lee et al. 2017). Infants and young children (U.S. Climate Change Science Program 2008), the elderly (Gamble et al. 2013), individuals with asthma or other respiratory conditions, outdoor workers, and low-income communities are more susceptible to heat-related illnesses, including heat cramps, heat exhaustion, heat rash, and heatstroke (Occupational Safety and Health Administration [OSHA], 1999).

   High temperatures frequently result in increased hospital visits or extended stays for individuals with underlying health conditions (U.S. Climate Change Science Program 2008; Vaidyanathan 2020). Between 2004 and 2018, the Centers for Disease Control and Prevention (CDC) recorded 10,527 heat-related deaths in the United States, averaging 702 deaths annually (Vaidyanathan et al. 2020). Similarly, across 93 European cities, heat islands are responsible for approximately 6,700 premature deaths each year, accounting for about 4.3% of total mortality during the summer months and 1.8% of year-round mortality.

6. **Heat Waves**: Heat waves cause heat exhaustion and heatstroke, which can be fatal. Rising temperatures are also associated with an increased risk of having a heart attack because the heart has to work harder to pump blood through vessels near the skin for cooling. For every one-degree increase in temperature, the risk of a heart attack rises by 1% (Phung et al. 2016).
7. **Effects on plant and animal life**: UHI has a significant impact on plant phenology (Zipper et al. 2016). The growing season of plants is lengthened, causing animal breeding seasons to extend (Intergovernmental Panel on Climate Change (IPCC)). This cycle exerts tremendous stress on the natural environment and makes it harder for most animals to survive the changes (Niemelä et al. 2011).
8. **Weather Conditions**: UHIs significantly impacts local weather patterns, with the average temperature rising during the day, reduced night-time cooling and the difference between day and night decreasing. It also increases the severity and duration of heat waves, putting vulnerable populations at risk. Increased clouds and fog, unusual wind patterns, and heavy downpours and thunderstorms are among the other impacts. Downwind of the UHI, weather patterns change, resulting in frequent precipitation each month, causing a dramatic shift in the ecosystem (collectiveweather.com; Siradel.com).
9. **Water bodies and aquatic life**: Heating up of infrastructures in UHIs, due to convection from rooftop structures and pavements, heats rainwater which drains into nearby water bodies like streams, rivers, and lakes (U.S. EPA 2024), causing a rise in urban water body temperatures. This negatively impacts aquatic animals, resulting in thermal shock (Briciu et al. 2020) affecting metabolism and reproduction, with rapid changes being especially harmful. Studies show that urban streams are hotter than those in forested areas, with temperatures rising by over 7°F during small storms due to heated urban runoff (Somers et al. 2016). Green infrastructure, such as rain gardens, permeable pavements, and green roofs, can help cool stormwater and improve water quality, offering a sustainable solution.
10. **Psychological and Mental Health Impacts:** UHIS intensify phychological stress by disrupting sleep, increasing cortisol levels, and worsening anxiety, irritability, and cognitive fatigue (Obradovich et al. 2017). Evidence shows that rising temperatures directly affect mental well-being: a major multi-decade study found that suicide rates increase by 0.7% in U.S. counties and 2.1% in Mexican municipalities for every 1 °C rise in monthly average temperature, with no signs of adaptation over time (Burke et al. 2018). The same study also reported increased depressive language in over 600 million social medias posts during hotter periods. Thus, UHIs contribute to significant mental health risks, including heightened stress, mood disturbances, and increased vulnerability among urban populations.

5 Conclusions

Urban Heat Islands are a pressing urban environmental issue with complex causes and significant impacts on health, ecosystems, and economies. Addressing UHIs requires an integrated approach involving urban planning, sustainable development, and community engagement. Strategies such as increasing green spaces, adopting cool roofing materials, and enhancing urban albedo can mitigate UHI effects. A deeper understanding of UHIs is crucial for building climate-resilient cities and ensuring sustainable urban living in the face of global warming.

References

Akbari H, Pomerantz M, Taha H (2001) Cool surfaces and shade trees to reduce energy use and improve air quality in urban areas. Sol Energy 70(3):295–310

Briciu AE, Mihăilă D, Graur A, Oprea DI, Prisăcariu A, Bistricean PI (2020) Changes in the water temperature of rivers impacted by the urban heat island: case study of Suceava city. Water 12(5):1343

Burke M, González F, Baylis P, Heft-Neal S, Baysan C, Basu S, Hsiang S (2018) Higher temperatures increase suicide rates in the United States and Mexico. Nat Clim Change 8(8):723–729. https://doi.org/10.1038/s41558-018-0222-x

Gamble JL, Hurley BJ, Schultz PA, Jaglom WS, Krishnan N, Harris M (2013) Climate change and older Americans: state of the science. Environ Health Perspect 121(1):15–22

Gill SE, Handley JF, Ennos AR, Pauleit S (2007) Adapting cities for climate change: the role of the green infrastructure. Built Environ 33(1):115–133

Grimmond SUE (2007) Urbanization and global environmental change: local effects of urban warming. Geogr J 173(1):83–88

https://collectiveweather.com/the-urban-heat-island-effect-how-cities-influence-local-weather/

https://www.siradel.com/urban-heat-island-effect-causes-and-solutions/

Lee YY, Din MFM, Ponraj M, Noor ZZ, Iwao K, Chelliapan S (2017) Overview of urban heat island (UHI) phenomenon towards human thermal comfort. Environ Eng Manag J (EEMJ)16(9)

Magura T, Ferrante M, Lövei GL (2020) Only habitat specialists become smaller with advancing urbanization. Glob Ecol Biogeogr 29(11):1978–1987

McMichael AJ, Woodruff RE, Hales S (2006) Climate change and human health: present and future risks. The Lancet 367(9513):859–869

Mogi M, Armbruster PA, Tuno N (2020) Differences in responses to urbanization between invasive mosquitoes, Aedes japonicus japonicu s (Diptera: Culicidae) and Aedes albopictus, in their native range, Japan. J Med Entomol 57(1):104–112

Niemelä J, Breuste JH, Guntenspergen G, McIntyre NE, Elmqvist T, James P (eds) (2011) Urban ecology: patterns, processes, and applications. OUP Oxford

Obradovich N, Migliorini R, Mednick SC, Fowler JH (2017) Nighttime temperature and human sleep loss in a changing climate. Sci Adv 3(5):e1601555. https://doi.org/10.1126/sciadv.1601555

Oke TR (1982) The energetic basis of the urban heat island. Q J R Meteorol Soc 108(455):1–24

OSHA (1999) Occupational safety and health administration, OSHA technical manual, Section III: chapter 4, heat stress. Occupational safety and health administration. https://www.osha.gov/dts/osta/otm/otm_iii/otm_iii_4.html

Parker DS, Barkaszi SF Jr (1997) Roof solar reflectance and cooling energy use: field research results from Florida. Energy Build 25(2):105–115

Phung D, Thai PK, Guo Y, Morawska L, Rutherford S, Chu C (2016) Ambient temperature and risk of cardiovascular hospitalization: an updated systematic review and meta-analysis. Sci Total Environ 550:1084–1102
Rizwan AM, Dennis LY, Chunho LIU (2008) A review on the generation, determination and mitigation of urban heat island. J Environ Sci 20(1):120–128
Salamanca F, Georgescu M, Mahalov A, Moustaoui M, Wang M (2014) Anthropogenic heating of the urban environment due to air conditioning. J Geophys Res: Atmos 119(10):5949–5965
Santamouris M (2014) On the energy impact of urban heat island and global warming on buildings. Energy Build 82:100–113
Santamouris M (2015) Analyzing the heat island magnitude and characteristics in one hundred Asian and Australian cities and regions. Sci Total Environ 512:582–598
Santamouris M (2020) Recent progress on urban overheating and heat island research. Integrated assessment of the energy, environmental, vulnerability and health impact. Synergies with the global climate change. Energy Build 207:109482
Somers KA, Bernhardt ES, Grace JB, Hassett BA, Sudduth EB, Wang S, Urban DL (2013) Streams in the urban heat island: spatial and temporal variability in temperature. Freshw Sci 32(1):309–326
Taha H (1997) Urban climates and heat islands: albedo, evapotranspiration, and anthropogenic heat. Energy Build 25(2):99–103
Urban Heat Islands 101—Resources for the Future. https://www.rff.org/publications/explainers/urban-heat-islands-101/
U.S. EPA (2008) Reducing urban heat islands: compendium of strategies. Environmental Protection Agency. https://www.epa.gov/heat-islands
U.S. Environmental Protection Agency (2008) Reducing urban heat islands: compendium of strategies. Draft. https://www.epa.gov/heat-islands/heat-island-compendium
U.S. EPA (2017) https://www.epa.gov/sites/default/files/2017-05/documents/reducing_urban_heat_islands_ch_1.pdf
U.S. EPA (2024) Urbanization and temperature. https://www.epa.gov/caddis/urbanization-temperature
Vaidyanathan A (2020) Heat-related deaths—united states, 2004–2018. MMWR. Morbidity and mortality weekly report 69
Villalobos-Jiménez G, Hassall C (2017) Effects of the urban heat island on the phenology of Odonata in London, UK. Int J Biometeorol 61:1337–1346
Zipper SC, Schatz J, Singh A, Kucharik CJ, Townsend PA, Loheide SP (2016) Urban heat island impacts on plant phenology: intra-urban variability and response to land cover. Environ Res Lett 11(5):054023

Historical Perspective: Urban Heat Islands (UHIs) Research and Its Significance

Bijeesh Kozhikkodan Veettil and Bhumika Das

Abstract As a result of rapid urbanization and infrastructure development, urban areas tend to have a lower albedo and higher land surface temperatures compared to their surrounding rural areas. This phenomenon is called the urban heat island (UHI) effect. Scientists have understood this concept since the early nineteenth century. The quantification of the UHI effect can be done either by measuring LST or atmospheric temperature. Most of the studies on historical development of UHIs have been done based on LST variations, air temperature measurements, and numerical modeling. The present study investigated the historical evolution of UHI research in terms of its definition, causes, effects, and mitigation. In addition, key developments in UHI research over time are also discussed.

Keywords Urban heat islands · Land surface temperature · Green structures · Urban microclimate · Thermal comfort

1 Introduction

More than 50% of the world's population, particularly in North America and Europe, lives in cities (Cohen 2003; Veettil and Van 2023). Even though this number is lower in Asia, it is reported that the urban population in the Asian region will increase by 64% by 2025 because of urban growth and industrialization (UN 2014, 2016). The rapid expansion of urban areas (cities, towns), also known as urban sprawl,

---

B. K. Veettil (✉)
Laboratory of Ecology and Environmental Management, Science and Technology Advanced Institute, Van Lang University, Ho Chi Minh City, Vietnam
e-mail: bkozhikkodanveettil@vlu.edu.vn

Faculty of Applied Technology, School of Technology, Van Lang University, Ho Chi Minh City, Vietnam

B. Das
Department of Civil Engineering, Lakhmi Chand Institute of Technology, Bilaspur, Chhattisgarh, India
e-mail: dr.bhumika.das@lcit.edu.in

B. Kozhikkodan Veettil and B. Das (eds.), *Remote Sensing of Urban Heat Islands*,
Cities and Nature, https://doi.org/10.1007/978-3-032-11580-5_2

has several negative environmental impacts (Veettil and Van 2023). These negative consequences of urban sprawl are deforestation, loss of habitat and biodiversity degradation, pollution, emission of greenhouse gases (CO_2, CH_4, N_2O, etc.), urban heat islands (UHI), and alterations in nutrient/mineral cycles (Estoque and Murayama 2014; Veettil and Grondona 2018; Veettil and Van 2023; Veettil et al. 2023). However, these negative impacts are caused by urban area expansion without proper planning (Ranagalage et al. 2014) and can be reduced by solutions, such as smart cities (Martis and Garau 2024) and sustainable cities (Petrucci 2023).

Several scientists have recently studied the UHI phenomenon among the above-mentioned environmental issues resulting from urban sprawl. The global mean temperature has increased due to industrialization and urbanization, particularly since the nineteenth century (Grondona et al. 2013). A UHI portion of a metropolitan area is marked by remarkably high temperatures, particularly at night with weak winds and during summer and winter peaks, compared to its rural areas. The UHI effect is an important environmental challenge all large and emerging urban areas face. Several UHI mitigation measures, including techniques to increase surface reflectance (albedo) and urban vegetation, can be achieved by a combination of measures, such as green pavements and roofs, urban forests, and reflective roofs (Hulley 2012).

2 Definition, Causes, and Effects of UHIs from a Historical Perspective

2.1 UHI—Definition and Historical Studies

According to Voogt and Oke (2003), UHI is the existence of higher atmospheric and land surface temperatures (LST) in the urbanized areas compared to the nearby rural areas. The intensity of UHI is a quantitative indicator of the thermal modifications imposed by a city upon the territory in which it is situated and of the relative warming compared to its rural areas at nighttime (Kim and Baik 2002; Memon et al. 2009; Martin-Vide et al. 2015). Even though the above-mentioned definition of UHI was published in the twentieth century, few researchers (e.g., Stewart 2019) identified the idea of UHI since the nineteenth century and over 40 studies were identified in the pre-1980 period. However, there is a lack of historical awareness in this subject (Stewart 2019). Many studies (e.g., Redway 1919; Sasakura 1931; Fukui and Wada 1941; Einarsson and Lowe 1955; Chandler 1960, 1965) in the 1900s and few before the twentieth century (e.g., Howard 1833; Renou 1868; Hann 1885) observed the differences in thermal comfort and temperature changes in urban areas before the term urban heat island came to existence in the late 1970s (e.g., Oke 1973, 1974, 1976). Nevertheless, the term urban heat island became popular after the publication of Oke (1973).

2.1.1 Quantification of UHI

Earlier observations on UHI have been made based on the air temperature measurements (e.g., Moreno 1895). According to Roth and Chow (2012), British scientist Luke Howard (1818) was (probably) the first to observe the differences in the air temperature between cities (London) and its rural surroundings. The early twentieth century measurements of temperature differences between urban and rural areas or between urban and suburban areas based on air temperature measurements in the United States were documented by few studies such as Hammon and Duenkel (1902) and Redway (1919). The first recorded urban–rural temperature measurement trip dates to the 1920s (Schmidt 1927).

The present-day quantification of UHI is mostly done based on measuring LST (using in situ, remote sensing and numerical modeling) (Veettil et al. 2023) or my air temperature measurements (Kolokotroni et al. 2007). Spatiotemporal measurement of LST from remote sensing data has been used as a reliable method for the quantification of UHI in several recent studies (Veettil and Van 2023; Veettil et al. 2023) whereas in situ observation in many cases can be inferior due to the sparse number of stations (Li et al. 2019). To overcome the temporal limitation of remote sensing observations and the spatial limitations of in situ measurements, studies, such as San Jose et al. (2013), recommended the use of numerical modeling to study and quantify UHIs.

2.2 UHI—Causes

Historically, the modifications in land surfaces from urbanization, particularly the replacement of vegetation cover by heat absorbing surfaces like concrete and asphalt, are the cause of UHIs. Several factors, varying from the presence of dark and impervious surfaces to the increase in greenhouse gases, influence the formation of UHI. Dark impervious surfaces, such as concrete, asphalt, and steel, absorb solar radiation and release heat back to the urban environments (Radhi et al. 2014). As a result, roads and pavements, railway lines and runways, and buildings' roofs in urban areas show higher LST than their surroundings. Jiang et al. (2018) considered roads (asphalt roads and runways in this context) as the major cause of the UHI effect. An example of LST distribution in urban areas (in Ho Chi Minh City, Vietnam) is given in Fig. 1.

Degradation and reduction in vegetated areas (forests, agriculture, grasslands) and waterbodies during rapid urban sprawl and unplanned urbanization also cause the formation of UHI and using concrete and asphalt is an inevitable component in building urban areas. Moreover, automobiles, factories, and air conditioners in urban areas cause increased heat, particularly due to aerosols and greenhouse gases.

It is widely observed that UHI is influenced and augmented by geometric effects. Tall buildings with multiple surfaces in urban areas cause increased reflection and absorption of sunlight, increasing heating. This effect is known as the urban canyon

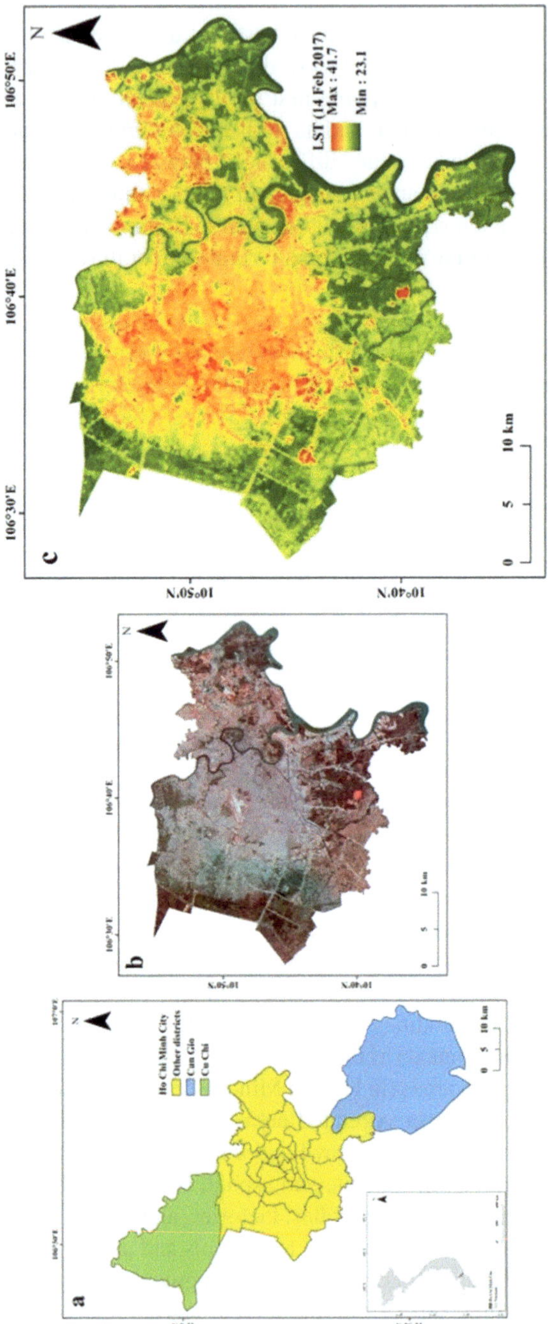

Fig. 1 LST distribution in Ho Chi Minh City (Vietnam)

effect. In addition, tall buildings block wind flow, thereby preventing convection cooling, and maintaining heat within the area.

2.3 *UHI—Effects*

The formation of UHI affects the environment, human and animal health, and quality of life in urban areas. Thermal discomfort is one of the direct effects of the UHI phenomenon. High energy demand (air conditioners, for example) is one of the economic aspects to reduce this thermal discomfort. This can lead to further air pollutants and greenhouse gas emissions, lower groundwater quality, and endangerment of urban biodiversity and human health (Jiang et al. 2018). Illness related to heat and thermal discomfort is common due to UHI, particularly in the summer months.

Several environmental issues were pointed out as influences of UHIs. Nevertheless, UHIs can have positive effects in cold climates that may reduce the demand for heating in such areas (Roth and Chow 2012). However, formation of UHI causes high thermal discomfort in tropical countries and results in high demand for residential cooling devices, such as air coolers. In addition to thermal discomfort, human health can be affected by UHIs in many ways, including exacerbating respiratory and cardiovascular diseases. High risk of heart-related illness is considered as a serious issue in urban areas (Cleland et al. 2023). Greenhouse gas emissions exacerbate the effects of UHIs (Veettil et al. 2023). Recent studies, such as Dai and Liu (2022), mentioned that, in addition to physical health, UHIs affect the mental health of urban residents.

Excess heat resulting from UHIs causes changes in flora and fauna in the urban ecosystems and air and water pollution (due to dust, microbes, etc.), particularly in tropical and subtropical climate regions (Roth and Chow 2012). The temperature variations alter aquatic ecosystems in urban areas, including the increase in invasive species (Cabon et al. 2023).

3 Mitigation of UHIs—Learning from the Past Toward a Sustainable Future

Mitigation of UHI effects can be achieved by strategies that can be applied in the project design phase of urban planning, thereby affecting the urban temperatures on a local scale (Gago et al. 2013). Such strategies include urban green spaces, specialized pavement surfaces and green roofs, and the use of distinct building types and materials (Gago et al. 2013).

Few studies, such as Gago et al. (2013) recommended modifications in urban morphology to alter certain parameters, such as green plot ratio, sky view factor, building density, wall surface area, pavement area, and albedo, to cause variations in energy consumption. Roth and Chow (2012) suggested reducing global carbon dioxide emissions by examining the footprint of human activity in urban areas as a part of sustainable city design. Three ways to achieve this, as given in Roth and Chow (2012), are (i) intelligent urban planning by encouraging competent usage of resources, (ii) building energy efficiency, and (iii) mitigating the UHI for reducing the use of air conditioning.

High-albedo roofing materials and pavements can be used to reduce heat absorption in urban areas. In addition, non-impervious pavements are also recommended in a few studies (e.g., Sailor 2006) to allow infiltration of water thereby cooling the pavement.

Even though some studies suggested urban waterbodies (e.g., ponds, urban lakes) as a UHI mitigation strategy (Robitu et al. 2006), some researchers do not agree with that as waterbodies have high heat absorption capacity and enhanced water surface cover tends to increase the UHI effect (Theuwes et al. 2012).

4 Conclusions

UHI is one of the negative effects of rapid urbanization and infrastructure development in urban areas that have lower albedos and higher land surface temperatures than their surrounding rural areas. Before the term urban heat island was introduced, the idea of higher temperatures or thermal discomfort in cities compared to their rural areas was understood since the early 1800s. The extent of the UHI effect and its quantification can be analyzed by measuring LST or atmospheric temperature or based on numerical modeling. Various factors causing the UHI effect include the presence of dark and impervious surfaces and the increase in greenhouse gas emissions. Degradation of vegetation coverage is pointed out as a major cause of UHI. In addition to thermal discomfort and higher energy demands for cooling, other negative effects of UHI in tropical and subtropical regions include heart and pulmonary diseases, air and water pollution, invasive species in aquatic environments, and alterations in flora and fauna in urban environments. Various methods to mitigate UHI include modifications in urban morphology to alter certain parameters to induce cooling effects, green surfaces, reducing carbon emissions, urban green fields, and sustainable urban planning.

References

Cabon V, Quénol H, Dubreuil V, Ridel A, Bergerot B (2023) Urban heat island and reduced habitat complexity explain spider community composition by excluding large and heat-sensitive species. Land 13(1):83. https://doi.org/10.3390/land13010083

Chandler TJ (1960) Wind as a factor of urban temperatures—a survey in north-east London. Weather 15:204–213

Chandler TJ (1965) The climate of London. Hutchinson, London

Cleland SE, Steinhardt W, Neas LM, West JJ, Rappold AG (2023) Urban heat island impacts on heat-related cardiovascular morbidity: a time series analysis of older adults in US metropolitan areas. Environ Int 178:108005. https://doi.org/10.1016/j.envint.2023.108005

Cohen JE (2003) Human population: the next half century. Science 302:1172–1175. https://doi.org/10.1126/science.1088665

Dai Y, Liu T (2022) Spatiotemporal mechanism of urban heat island effects on human health—evidence from Tianjin city of China. Front Ecol Evol 10:1010400. https://doi.org/10.3389/fevo.2022.1010400

Einarsson E, Lowe AB (1955) A study of horizontal temperature variations in the Winnipeg area on nights favouring radiational cooling. Canadian Department of Transport, Meteorological Division, CIR-2647, TEC-214, Toronto

Estoque RC, Murayama Y (2014) Measuring sustainability based upon various perspectives: a case study of a hill station in Southeast Asia. Ambio 43:943–956. https://doi.org/10.1007/s13280-014-0498-7

Fukui E, Wada T (1941) Horizontal distribution of air temperature in the great cities of Japan. Japanese Geogr Rev Jpn 17:354–372

Gago EJ, Roldan J, Pachero-Torres R, Ordonez J (2013) The city and urban heat islands: a review of strategies to mitigate adverse effects. Renew Sustain Energy Rev 25:749–758. https://doi.org/10.1016/j.rser.2013.05.057

Grondona AEB, Veettil BK, Rolim SBA (2013) Urban heat island development during the last two decades in Porto Alegre, Brazil, and its monitoring. In: Proceedings of the joint urban remote sensing event (JURSE)-2013, Sao Paulo, Brazil, pp 61–64. https://doi.org/10.1109/JURSE.2013.6550666

Hammon WH, Duenckel FW (1902) Abstract of a comparison of the minimum temperatures recorded at the United States weather Bureau and the forest park meteorological observatories, St. Louis, MO., for the year 1891. Mon Weather Rev 30:12–13

Hann J (1885) Über den Temperaturunterschied zwischen Stadt und Land [On the temperature difference between town and country]. Österreichischen Gesellschaft Für Meteorologie, Zeitschrift 20:457–462

Howard L (1833) The climate of London deduced from meteorological observations, made in the metropolis, and at various places around it vols 1–3, Harvey and Darton, London

Hulley ME (2012) 5—The urban heat island effect: causes and potential solutions. In: Meman F (ed) Metropolitan sustainability understanding and improving the urban environment, pp 79–98. https://doi.org/10.1533/9780857096463.1.79

Jiang W, Huang Y, Sha A (2018) A review of eco-friendly functional road materials. Constr Build Mater 191:1082–1092. https://doi.org/10.1016/j.conbuildmat.2018.10.082

Kim YH, Baik JJ (2002) Maximum urban heat island intensity in Seoul. J Appl Meteorol 41:651–659. https://doi.org/10.1175/15200450(2002)041%3c0651:MUHIII%3e2.0.CO;2

Kolokotroni M (2007) The London heat island and building cooling design. Sol Energy 81:102–110

Li H, Zhou Y, Wang X, ZhouX ZH, Sodiudi S (2013) Quantifying urban heat island intensity and its physical mechanism using WRF/UCM. Sci Total Environ 650:3110–3119. https://doi.org/10.1016/j.scitotenv.2018.10.025

Martin-Vide J, Sarricolea P, Moreno-García MC (2015) On the definition of urban heat island intensity: the "rural" reference. Front Earth Sci 3:24. https://doi.org/10.3389/feart.2015.00024

Memon RA, Leung DYC, Liu CH (2009) An investigation of urban heat island intensity (UHII) as an indicator of urban heating. Atmos Res 94:491–500. https://doi.org/10.1016/j.atmosres.2009.07.006

Moreno M (1895) Comparación de los climas de México y Tacubaya [A comparison between the climates of Mexico City and Tucabaya]. In: Memorias de la Sociedad Científica "Antonio Alzate," Tomo IX. Imprenta del Gobierno Federal, México

Mutzu Martis M, Garau C (2024) A literature review of the urban heat island (UHI) phenomenon connected with smart cities paradigm. In: Gervasi O, Murgante B, Garau C, Taniar DC, Rocha

AMA, Faginas Lago MN (eds) Computational science and its applications—ICCSA 2024 workshops. ICCSA 2024. Lecture Notes in Computer Science, vol 14823. Springer, Cham. https://doi.org/10.1007/978-3-031-65329-2_1
Oke TR (1973) City size and the urban heat island. Atmos Environ 1967(7):769–779. https://doi.org/10.1016/0004-6981(73)90140-6
Oke TR (1976) The distinction between canopy and boundary-layer urban heat islands. Atmosphere 14:269–277. https://doi.org/10.1080/00046973.1976.9648422
Oke TR (1974) A review of urban climatology 1968–1973. WMO Technical Note No. 134, WMO No. 383. World Meteorological Organization, Geneva
Petrucci AL (2024) Environmental retrofitting, fighting urban heat island toward NEZ sustainable smart cities. In: Belaïd F, Arora A (eds) Smart cities. Studies in energy, resource and environmental economics. Springer, Cham. https://doi.org/10.1007/978-3-031-35664-3_8
Radhi H, Assem E, Sharples S (2014) On the colours and properties of building surface materials to mitigate urban heat islands in highly productive solar regions. Build Environ 72:162–172. https://doi.org/10.1016/j.buildenv.2013.11.005
Ranagalage M, Estoque RC, Murayama Y (2014) An urban heat island study of the Colombo metropolitan area, Sri Lanka, based on Landsat data (1997–2017). Int J Geo-Inf 6(7):189. https://doi.org/10.3390/ijgi6070189
Redway JW (1919) Urban versus suburban temperatures. Monthly Weather Rev 47:28–29
Renou E (1868) Différences de température entre la ville et la campagne [Temperature differences between town and country]. Annuaire, Société Météorologique De France 16:83–97
Robitu M, Musy M, Inard C, Groleau D (2006) Modeling the influence of vegetation and water pond on urban microclimate. Sol Energy 80(4):435–447
Roth M, Chow WTL (2012) A historical review and assessment of urban heat island research in Singapore. Singap J Trop Geogr 33:381–397. https://doi.org/10.1111/sjtg.12003
Sailor DJ (2006) Mitigation of urban heat islands—recent progress and future prospects. Paper presented at the Paper presented on American meteorological society 6th symposium on the urban environment and forum on managing our physical and natural resources
San José R, Pérez JL, González RM (2013) Very high resolution urban simulations with WRF/UCM and CMAQ over European cities. Urban Environment. Springer, Dordrecht, pp 293–301
Sasakura K (1931) Air temperature difference between Tokyo and its suburb. (In Japanese). Geogr Rev Jpn 7:29–37
Schmidt W (1927) Die Verteilung der Minimumtemperaturen in der Frostnacht des 12. Mai 1927 im Gemeindegebiete von Wien [Distribution of minimum tem peratures during the frost night of May 12, 1927, within the communal limits of Vienna]. Fortschritte der Landwirtschaft 2:681–686
Stewart ID (2019) Why should urban heat island researchers study history? Urban Climate 30:100484. https://doi.org/10.1016/j.uclim.2019.100484
Theeuwes N, Steenevel G, Ronda R, Heusinkveld B, Holtslag A (2012) 197: Mitigation of the urban heat island effect using vegetation and water bodies. ICUC8—8th International conference on urban climates, 6th–10th August, 2012, UCD, Dublin, Ireland
UN (2016) The world's cities in 2016–data booklet (ST/ESA/SER.A/ 392). Department of economic and social affairs, population division, United Nations, New York, p 29
United Nations (UN) (2014) World urbanization prospects: the 2014 revision—highlights. United Nations, New York
Veettil BK, Grondona AEB (2018) Vegetation changes and formation of small-scale urban heat islands in three populated districts of Kerala State, India. Acta Geophys 66:1063–1072. https://doi.org/10.1007/s11600-018-0189-z
Veettil BK, Puri V, Van DD, Quang NX (2023) Variations in land surface temperatures in small-scale urban areas in Vietnam during Covid-19 restrictions: case studies from Da Nang, Hue and Vinh City. Environ Monit Assess 195:822. https://doi.org/10.1007/s10661-023-11332-4
Veettil BK, Van DD (2023) Did the Covid-19 restrictions influence land surface temperatures in Southeast Asia? A study from Ho Chi Minh City, Vietnam. Environ Sci Pollut Res 30:66812–66821 (2023). https://doi.org/10.1007/s11356-023-26892-8

Voogt JA, Oke TR (2003) Thermal remote sensing of urban climates. Remote Sens Environ 86:70–384. https://doi.org/10.1016/S0034-4257(03)00079-8

Technologies and Instruments: Satellites and Sensors for Urban Heat Island Research

Bijeesh Kozhikkodan Veettil and Bhumika Das

Abstract Increased thermal discomfort is a byproduct of urbanization and expansion of impervious surfaces. Urban Heat Islands (UHIs) have been the core topic of many researchers in the past few decades. Several studies have used remote sensing data to understand the formation and expansion of UHIs. Land surface temperature (LST) measurements have been used as a proxy in the majority of the UHI studies using remote sensing. Several satellite sensors have been launched since the 1960s and those having thermal wavelengths (3 μm–14 μm) have been used for estimating LSTs in general and urban climatology in particular. This chapter discussed various satellites and sensors that have helped the scientific community in advancing the study of UHIs globally.

Keywords Urban heat islands · Land-surface temperature · Satellite remote sensing · Thermal imagery · Thermal mapping

1 Introduction

The explosion of population and rapid urbanization (or urban sprawl) have caused several threats to the environment (Veettil and Van 2023). Some of these key threats were listed as deforestation, loss of habitat and biodiversity degradation, pollution, emission of greenhouse gases, urban heat islands (UHI), and alterations in nutrient/mineral cycles (Veettil and Van 2023; Veettil et al. 2023). Nevertheless, methods to

B. K. Veettil (✉)
Laboratory of Ecology and Environmental Management, Science and Technology Advanced Institute, Van Lang University, Ho Chi Minh City, Vietnam
e-mail: bkozhikkodanveettil@vlu.edu.vn

Faculty of Applied Technology, School of Technology, Van Lang University, Ho Chi Minh City, Vietnam

B. Das
Department of Civil Engineering, Lakhmi Chand Institute of Technology, Bilaspur, Chhattisgarh, India
e-mail: dr.bhumika.das@lcit.edu.in

B. Kozhikkodan Veettil and B. Das (eds.), *Remote Sensing of Urban Heat Islands*, Cities and Nature, https://doi.org/10.1007/978-3-032-11580-5_3

monitor urban sprawl can help mitigate such negative impacts on the environment. Satellite remote sensing has been helpful to understand the spatiotemporal changes in land use and land cover (LULC), land-surface temperatures (LST) and formation of UHIs.

Formation and expansion of UHIs in rapidly growing urban areas have attracted the global scientific community in recent decades, even though the idea of urban areas having a higher thermal discomfort compared to their surroundings dates to the early nineteenth century (Stewart 2019). UHI can be defined as the existence of higher atmospheric and land-surface temperatures (LST) in the urbanized areas compared to the nearby rural areas (Voogt and Oke 2003). The presence of UHIs can be identified based on the air temperature or land-surface temperature measurements (Veettil et al. 2023) and several studies (e.g., San Jose et al. 2013) recommended using numerical studies for understanding the expansion of UHIs.

Studies on UHIs, particularly those based on the measurement of LST, have gained momentum since the launch of satellites for land cover monitoring in the mid-1900s. The launch of sensors that can capture information in thermal infrared wavelengths (between 3 μm and 14 μm) has revolutionized the study of UHIs (Weng 2009). Various satellite sensors that can collect thermal information include Heat Capacity Mapping Mission (HCMM), Landsat TM/ETM + /TIRS, Advanced very-high-resolution radiometer (AVHRR), Moderate Resolution Imaging Spectroradiometer (MODIS), Advanced Spaceborne Thermal Emission and Reflection Radiometer (ASTER), Thermal Infrared Multispectral Scanner (TIMS) (Weng 2009), and Television Infrared Observation Satellite (TIROS-II) (Prata et al. 1995).

2 Thermal Remote Sensing and Its Use in UHI Research

Thermal remote sensing essentially deals with acquiring, processing, and interpreting data acquired in the thermal infrared region of the electromagnetic spectrum. Instead of measuring the reflected radiation from the target surface, thermal remote sensing measures the radiation emitted from the surface (Singh et al. 2024). Essentially all natural surfaces emit thermal radiation and sensors can detect this property and understand land-surface properties, particularly by measuring the surface temperature. Therefore, thermal remote sensing is a valuable tool in UHI research in high spatial and temporal resolution (Singh et al. 2024).

The basic principle exploited by thermal remote sensing is that any surface above absolute zero (0 K or −273.15 °C) emits radiation in the infrared range of the electromagnetic spectrum. The quantity of radiated energy and the wavelength of emitted radiation depends on the surface emissivity (ε) and kinetic temperature (usually measured in K or °C). Unlike optical data, thermal information acquired during the nighttime is preferred by many researchers as the effect of solar heating will be minimal.

Thermal remote sensing is considered as a relatively different field of remote sensing compared to those that use optical and microwave regions. In many cases, thermal data provides complementary information to other remote sensing information, and its full potential is still not used in geomorphological applications.

LST is a key variable essential to model the surface energy budget accurately. LST is used to parametrize the two-way energy transfers between the atmosphere and surface in models (Prata et al. 1995). Despite various weather forecasting models being capable of estimating surface temperatures, satellite-derived LST values provide accurate and useful input in such models. Moreover, LST derived from satellites is used for many applications, including regional evapotranspiration estimation, frost detection and forecasting, monitoring water stress in crops, and thermal inertia studies (Prata et al. 1995).

In many cases, satellite-derived LST does not consider atmospheric effects in the imagery, and the lack of atmospheric correction implies that the algorithm considered atmospheric effects to be the same in the entire area (Rasul et al. 2017). However, water vapor and pollutant contents vary horizontally in urban areas which is a source of error in the estimated LST (Rasul et al. 2017). Another source of errors in urban heat island studies using satellite data is due to the seasonality or unknown variations in temperature or atmospheric effects rather than the development of UHI. Estimating a high temperature from a satellite image compared to another one does not mean the temperature has increased and it is possible that at the image acquisition time, the temperature was high for other reasons (Rasul et al. 2017).

3 Thermal Sensors Onboard Satellites

The measurement of land-surface temperatures based on thermal remote sensing from satellites began in the early 1960s with the launch of the TIROS-II satellite (Prata et al. 1995), proceeded by AVHRR by the National Oceanic and Atmospheric Administration (NOAA). The AVHRR on board the NOAA series of polar-orbiting satellites has been used for several applications including LST.

TIROS-II imagery has been used to measure LST in clear, dry land areas (Rao and Winston 1963). However, this data fails if cirriform cloudiness is present (e.g., in the Gulf of Mexico). Moreover, the accuracy of LST from this data is limited in areas where field temperature measurements are not available for estimating the accuracy of satellite-derived LST.

AVHRR imagery has been used for LST measurements and UHI research based on split-window techniques in several studies (e.g., Lee 1993; Stathopoulou and Cartalis 2009). Even though the spatial resolution is limited, urban land use, such as built-up, residential, and industrial areas, can be identified from the AVHRR thermal data (Lee 1993). Moreover, Stathopoulou and Cartalis (2009) observed the spatial pattern of the downscaled AVHRR LST is visually better than the original AVHRR LST and resembles that derived from Landsat.

Landsat series (TM, ETM + , and TIRS) imagery has a long history of UHI research and satellite-derived LST estimation (Veettil and Grondona 2018). Landsat TM data is one of the most widely used satellite images for estimating LST because of its high resolution (120 m) (compared to AVHRR or MODIS) and its free availability from the US Geological Survey (USGS), which has one thermal infrared (TIR) band (Liu and Zhang 2011). Its helio-synchronous orbit favors temporal studies of the same place since its revisit occurs at the same point and UTM time, with an interval of 16 days (Almeida et al. 2021).

ASTER was launched in December 1999 onboard NASA's Terra platform. The ASTER Level 1 Precision Terrain Corrected Registered At-Sensor Radiance (AST_L1T) data contains calibrated at-sensor radiance (Rolim et al. 2020). This multi-file product includes data in visible and near-infrared, shortwave infrared, and thermal infrared channels with spatial resolutions of 15, 30, and 90 m, respectively (Rolim et al. 2020). ASTER data has been used by several studies on UHI monitoring as five thermal bands are available (most of the studies used Band 13 as other bands are found to be influenced by atmospheric noise) (Nichol et al. 2009). Most of these studies used the Temperature-Emissivity Separation (TES), Reference Channel Method (RCM), or Emissivity Normalization Method (ENM) when using ASTER data for estimating LST (Rolim et al. 2020). In addition, several studies (e.g., Cai et al. 2011) combined Landsat and ASTER data in UHI research with improved accuracy or resolution.

Launched in 1978, Heat Capacity Mapping Mission (HCMM) data has a wavelength range of 10.2–12.5 μm. The HCMM satellite acquires visible and infrared data at a spatial resolution of 500 m from 620 km altitude and, despite the data transmission being analogous to ground stations, a high-performance design has demonstrated a temperature data delivering capacity with a noise equivalent temperature of 0.4 K (Price 1979). This data has been used together with Landsat MSS data (digitally integrated) to improve land-surface classification (Witt et al. 1985) as well as to understand the presence and microstructure of UHIs. The minimum air temperatures averaged within climatically similar regions are found to be linearly related to HCMM-derived night-time LST (Byrne et al. 1984).

MODIS is an instrument onboard Terra (1999) and Aqua (2002) satellites, launched by NASA. MODIS thermal data has a resolution of 1 km, which makes its products applicable to larger regions. In MODIS products, the LST value is retrieved by applying the generalized split-window algorithm. TIR sensor of the MODIS measures the top of the atmosphere (TOA) radiation from which the brightness temperature can be estimated using Planck's law (Monteiro et al. 2021). However, due to the satellite's non-vertical viewing angle, urban geometry, variable surface emissivity, and various atmospheric effects, there will be a difference of 1 K–5 K in the estimated brightness temperature from the real LST (Dousset and Gourmelon 2003), which can be removed based on various correction algorithms (Monteiro et al. 2021).

TIMS measures the radiation from the surface modified by the atmosphere in six thermal channels located (8 μm and 12.5 μm) from which LST can be extracted using a variety of algorithms, such as Temperature-Emissivity Separation (TES) (Kealy and Hook 1993; Schmugge et al. 1998). It was observed that soil moisture content may affect the accuracy (by 2–3%) of LST estimated using TIMS data (Schmugge et al. 1998).

4 Conclusions

This chapter provided a brief discussion of different satellite sensors that have been used in UHI research. Thermal sensors operate in the 8–14 μm range of the electromagnetic spectrum, which includes the emission part of the earth's spectrum, making heat detection by RS sensors possible. Several satellite data, such as HCMM, Landsat TM/ETM + /TIRS, AVHRR, MODIS, ASTER, TIMS, and TIROS-II, are available for time-series analysis of UHI formation and development globally. Different data have constraints in spatial resolution, spectral range, and coverage, which can be overcome by integrating data from multiple sensors.

References

Almeida CRD, Teodoro AC, Gonçalves A (2021) Study of the urban heat island (UHI) using remote sensing data/techniques: a systematic review. Environments 8(10):105. https://doi.org/10.3390/environments8100105

Byrne GF, Kaima JD, Streten NA (1984) On the relation between HCMM satellite data and temperatures from standard meteorological sites in complex terrain. Int J Remote Sens 5:56–77. https://doi.org/10.1080/01431168408948789

Cai G, Du M, Xue Y (2011) Monitoring of urban heat island effect in Beijing combining ASTER and TM data. Int J Remote Sens 32:1213–1232. https://doi.org/10.1080/01431160903469079

Dousset B, Gourmelon F (2003) Satellite multi-sensor data analysis of urban surface temperatures and landcover. ISPRS J Photogramm Remote Sens 58:43–54. https://doi.org/10.1016/S0924-2716(03)00016-9

Kealy PS, Hook SJ (1993) Separating temperature and emissivity in thermal infrared multispectral scanner data: implications for recovering land surface temperatures. IEEE Trans Geosci Remote Sens 31:1155–1164. https://doi.org/10.1109/36.317447

Lee H-Y (1993) An application of NOAA AVHRR thermal data to the study of urban heat islands. Atmospheric Environment. Part b. Urban Atmosphere 27:1–13. https://doi.org/10.1016/0957-1272(93)90041-4

Liu L, Zhang Y (2011) Urban heat island analysis using the landsat TM data and ASTER data: a case study in Hong Kong. Remote Sens 3(7):1535–1552. https://doi.org/10.3390/rs3071535

Monteiro FF, Goncalves WA, Andrade LMB, Villavencio LMM, Silva CMS (2021) Assessment of urban heat islands in Brazil based on MODIS remote sensing data. Urban Clim 35:100726. https://doi.org/10.1016/j.uclim.2020.100726

Nichol JE, Fung WY, Lam K, Wong MS (2009) Urban heat island diagnosis using ASTER satellite images and 'in situ' air temperature. Atmos Res 94:276–284. https://doi.org/10.1016/j.atmosres.2009.06.011

Prata AJ, Caselles V, Coll C, Sobrino JA, Ottle C (1995) Thermal remote sensing of land surface temperature from satellites: current status and future prospects. Remote Sens Rev 12:175–224. https://doi.org/10.1080/02757259509532285

Price JC (1979) Assessment of the urban heat island effect through the use of satellite data. Mon Weather Rev 107:1554–1557. https://doi.org/10.1175/1520-0493(1979)107%3c1554:AOTUHI%3e2.0.CO;2

Rao P, Winston JS (1963) An investigation of some synoptic capabilities of atmospheric "window"measurements from satellite TIROSII. Appl Meteorol 2:12–23

Rasul A, Balzter H, Smith C, Remedios J, Adamu B, Sobrino JA, Srivanit M, Weng Q (2017) A review on remote sensing of urban heat and cool islands. Land 6(2):38. https://doi.org/10.3390/land6020038

Rolim SBA, Veettil BK, Kafer PS, Grondona AEB, Igleias ML, Diaz LR, Hackmann CL (2020) Comparison of emissivity retrieval methods from ASTER data using fourier-transform infrared spectroscopy. Acta Geophys 68:1867–1876. https://doi.org/10.1007/s11600-020-00498-x

San José R, Pérez JL, González RM (2013) Very high-resolution urban simulations with WRF/UCM and CMAQ over European cities. Urban Environment. Springer, Dordrecht, pp 293–301

Schmugge T, Hook SJ, Coll C (1998) Recovering surface temperature and emissivity from thermal infrared multispectral data. Remote Sens Environ 65:121–131. https://doi.org/10.1016/S0034-4257(98)00023-6

Singh S, Mall RK, Chaturvedi A, Singh N, Srivastava PK (2024) Chapter 6—Advances in remote sensing in measuring urban heat island effect and its management. In: Kumar A, Srivastava PK, Saikia P, Mall RK (eds) Earth Observation in urban monitoring—techniques and challenges, Elsevier, pp 113–132. https://doi.org/10.1016/B978-0-323-99164-3.00011-2

Stathopoulou M, Cartalis C (2009) Downscaling AVHRR land surface temperatures for improved surface urban heat island intensity estimation. Remote Sens Environ 113:2592–2605. https://doi.org/10.1016/j.rse.2009.07.017

Stewart ID (2019) Why should urban heat island researchers study history? Urban Clim 30:100484. https://doi.org/10.1016/j.uclim.2019.100484

Veettil BK, Grondona AEB (2018) Vegetation changes and formation of small-scale urban heat islands in three populated districts of Kerala State, India. Acta Geophys 66:1063–1072. https://doi.org/10.1007/s11600-018-0189-z

Veettil BK, Puri V, Van DD, Quang NX (2023) Variations in land surface temperatures in small-scale urban areas in Vietnam during Covid-19 restrictions: case studies from Da Nang, Hue and Vinh City. Environ Monit Assess 195:822. https://doi.org/10.1007/s10661-023-11332-4

Veettil BK, Van DD (2023) Did the Covid-19 restrictions influence land surface temperatures in Southeast Asia? A study from Ho Chi Minh City, Vietnam. Environ Sci Pollut Res 30:66812–66821 (2023). https://doi.org/10.1007/s11356-023-26892-8

Voogt JA, Oke TR (2003) Thermal remote sensing of urban climates. Remote Sens Environ 86:70–384. https://doi.org/10.1016/S0034-4257(03)00079-8

Weng Q (2009) Thermal infrared remote sensing for urban climate and environmental studies: methods, applications, and trends. ISPRS J Photogramm Remote Sens 64:335–344. https://doi.org/10.1016/j.isprsjprs.2009.03.007

Witt RG, Minor TB, Sekhon RS (1985) Use of HCMM thermal data to improve accuracy of MSS land-surface classification mapping. Int J Remote Sens 6:1623–1636. https://doi.org/10.1080/01431168508948310

Thermal Remote Sensing: A Case Study on Detecting and Analyzing Surface Temperature Variations in Response to Land Use and Land Cover Changes

Lucas Lima Raiol, Dayla Carolina Rodrigues Santos, Maria Luiza Nascimento Dias, João Fernandes da Silva Júnior, Sanae Nogueira Hayashi, Luiz Cláudio Moreira Melo Júnior, Nelton Cavalcante da Luz, and Andrés Velastegui-Montoya

Abstract This study analyzed variations in Land Surface Temperature (LST) and its relationship with changes in Land Use and Land Cover (LULC) in the municipality of Capanema, Eastern Amazon. The research utilized thermal imagery from Landsat satellites (TM and OLI/TIRS) to extract LST and Normalized Difference Vegetation Index (NDVI) values from 1985 to 2023. LULC classification was obtained from *MapBiomas* data, enabling the identification of landscape changes over time. Statistical analyses, including Spearman's correlation, Kruskal–Wallis, and Dunn's Post Hoc tests, were conducted to assess the relationship between LST, NDVI, and LULC. The results revealed a significant temperature increase over the period analyzed, with an overall rise of 4.7 °C. Urban and pasture areas exhibited the highest LST values, while forested areas demonstrated a strong cooling effect. A negative correlation was found between NDVI and LST, confirming the role of vegetation in regulating surface temperatures. These findings highlight the need for mitigation and adaptation strategies, including sustainable urban planning and reforestation initiatives, to address rising temperatures. Furthermore, they emphasize the importance of remote sensing and statistical analyses for environmental monitoring, providing critical insights for

L. L. Raiol (✉) · D. C. R. Santos · M. L. N. Dias
Institute of Geosciences (IG), Federal University of Pará (UFPA), Belém, Brazil
e-mail: lucas.raiol@ig.ufpa.br

J. F. da Silva Júnior · S. N. Hayashi · L. C. M. M. Júnior
Federal Rural University of Amazonia (UFRA), Belém, Brazil

N. C. da Luz
National Institute for Space Research (INPE), Belém, Brazil

A. Velastegui-Montoya
Laboratory of Geoinformation and Remote Sensing, ESPOL Polytechnic University, Guayaquil, Ecuador

Faculty of Engineering in Earth Sciences, ESPOL Polytechnic University, Guayaquil, Ecuador

B. Kozhikkodan Veettil and B. Das (eds.), *Remote Sensing of Urban Heat Islands*, Cities and Nature, https://doi.org/10.1007/978-3-032-11580-5_4

public policies aimed at conservation and sustainable development in the Eastern Amazon.

Keywords Land surface temperature · Urban heat island effect · Land use and land cover changes · Amazon

1 Introduction

Human activities are estimated to have caused about 1.0 °C of global warming above pre-industrial levels (1850–1900), ranging from 0.8 °C to 1.20 °C (IPCC 2023). Global warming was expected to reach 1.5 °C between 2030 and 2052 (IPCC 2023). However, the year 2024 exceeded 1.5 °C, being mainly human-induced, in which extreme air and sea surface temperatures (SST) are influenced, while other factors, such as the El Niño Southern Oscillation (ENSO), also contributed to the unusual temperatures observed during the year (Copernicus 2025; IPCC 2023).

In the Amazon basin, the impacts of unprecedented droughts were experienced in the austral summer of 2022–2023 and extending into 2024, with negative rainfall anomalies (−50 mm to 100 mm/month), positive air temperatures (+3 °C) including four intense heat waves during the dry-to-wet transition season, September–October–November (SON) 2023, particularly in the eastern Amazon (Espinoza et al. 2024; Marengo et al. 2024). Warm sea surface temperature (SST) anomalies due to El Niño and the warm tropical North Atlantic started in May 2023. They intensified in SON 2023, helping to induce these changes (Marengo et al. 2024). During this period, heat waves were reported in the Amazon, causing changes to aquatic habitats, an increase in the incidence of fire, and a decrease in river levels (Espinoza et al. 2024; Fleischmann et al. 2024; Marengo et al. 2024; Moreira et al. 2024; Raiol et al. 2024).

In this sense, the increase in land surface temperature (LST) is widely addressed in studies of urban heat islands (UHI) in response to land use and land cover (LULC), in which urban expansion and vegetation decrease tend to increase LST compared to rural areas (Furtado et al. 2024; Kafy et al. 2022; Rahaman et al. 2022; Raiol et al. 2024; Waleed et al. 2023). Urban areas characterized by built-up surfaces like buildings and roads tend to have higher LSTs than landscapes such as forests and water bodies. This is due to the heat-retaining properties of urban materials and the reduction of vegetation, which otherwise provides cooling through evapotranspiration (Bokaie et al. 2016; Kafy et al. 2020).

Thermal remote sensing is a critical tool for monitoring LST, a key parameter in understanding land–atmosphere interactions. LST is influenced by various factors, including atmospheric conditions, land cover, and surface properties, and is essential for applications in climate change studies, urban planning, and environmental monitoring (Almeida et al. 2021; Diem et al. 2024; Raiol et al. 2024). The use of thermal sensors from the MODIS and Landsat satellites stands out, the latter being the most widespread in the world, as it is appropriate for long-term monitoring, while MODIS provides a daily to monthly analysis (Almeida et al. 2021; Diem et al. 2024; Raiol et al. 2024).

Retrieving accurate LST from thermal infrared (TIR) remote sensing data involves several challenges. Key error sources include atmospheric effects, land surface emissivity, and sensor noise (Malakar et al. 2018). Algorithms implemented for LST retrieval involving atmospheric correction using a radiative transfer model and reanalysis data, followed by emissivity adjustments and a lookup-table approach for LST retrieval, presented Root Mean Square Error (RMSE) that ranged from the mean bias of 0.7 K and RMSE of 2.2 K for specific sites (Malakar et al. 2018). While the single-channel framework implemented on Google Earth Engine unifies LST, land surface emissivity, and atmospheric water vapor estimation achieved an average bias of 0.33 K and RMSE of 2.01 K, also indicating good agreement between Landsat and MODIS LST retrievals (Wang et al. 2020). In the municipality of Belém, in the eastern Amazon, an algorithm implemented using LST retrieval from Landsat Collection 2 calibrated top-of-atmosphere (TOA) reflectance found good agreement within situ station mediation with a strong positive correlation ($\rho = 0.99$). The coefficient of determination explained 97% of the variation in the LST data, and the calculated RMSE was 0.59 °C (Raiol et al. 2024).

Recent advancements in LST retrieval algorithms have improved accuracy under various conditions. For instance, mid-infrared (MIR) channels perform better than TIR channels under humid conditions due to less water vapor absorption, leading to more stable LST retrievals (Ye et al. 2024). Moreover, combining TIR and passive microwave (PMW) data allows for all-weather LST retrieval, overcoming the limitations of cloud cover in TIR measurements (Duan et al. 2017).

LST data derived from remote sensing have wide-ranging applications. They are crucial for monitoring UHI, as demonstrated by studies using high-resolution drone imagery to capture spatial variability in urban environments (Naughton and McDonald 2019). LST variations are also linked to vegetation indices, providing insights into urban expansion and its thermal impacts (Yang et al. 2020). Furthermore, LST data are used in soil moisture estimation, which is vital for understanding hydrological processes (Zhang et al. 2014).

In the context of the municipality of Capanema, there is a gap in the use of remote sensing to analyze LULC and LST to understand their intrinsic relationship, as well as the importance of the municipality of Capanema in the regional scenario because it is a medium-sized city. Thus, the objective was to analyze the LST and its relationship with LULC in the municipality of Capanema, in the eastern Amazon, through Landsat thermal sensors to understand the relationship between LST variations in response to LULC.

2 Material and Methods

2.1 Study Area

The municipality of Capanema (Fig. 1) is one of the most populous cities in the northeast region of Pará State, in the North of Brazil, with 70,394 inhabitants and a territory of 621,483 km^2 (IBGE 2022). Its borders are to the north with the municipalities of Primavera and Quatipuru, to the east with Tracuateua and Santa Luzia do Pará, to the south with Ourém and Bonito, and the west with Bonito and Peixe-Boi. The municipality of Capanema plays a fundamental role in the northeast of Pará, as it has a dynamic trade in public and private services, in addition to its privileged geographical position in the road network, the main road being BR-163 (Raiol et al. 2019; Ribeiro 2018).

According to the Köppen climate classification, the municipality of Capanema corresponds to the Am categories (monsoon) with three dry months, characterized by an annual average rainfall of around 2,200 mm, high air humidity almost all year round, and high temperatures with an annual average of 26 °C (Alvares et al. 2013; Fapespa 2024). The vegetation in the municipality is predominantly characterized by dense rainforest, which experiences periods of intense and continuous rainfall. This vegetation comprises species with extensive, evergreen leaves and is primarily found in alluvial subformations. Additionally, riparian and floodplain vegetation is

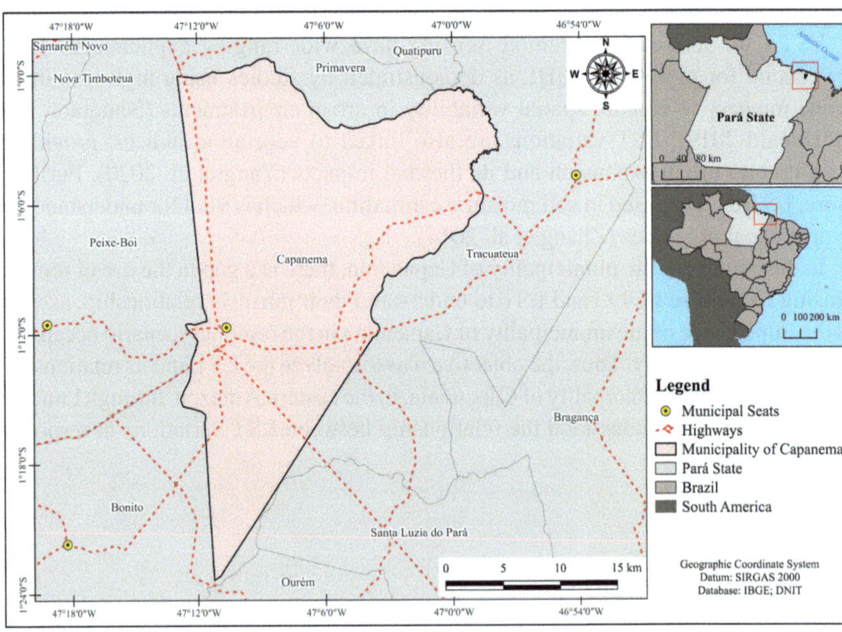

Fig. 1 Location of the municipality of Capanema

present, occupying riverbanks and areas subject to periodic flooding (Fapespa 2024). The predominant soil classes in the municipality include medium-textured yellow Oxisols, Entisols, hydromorphic soils without clear differentiation, and alluvial soils, often found in associations, as well as quartz sand (Fapespa 2024).

2.2 *Research Framework*

The methodological procedures involved three phases: (I) Data Processing, involving the acquisition and retrieval of the LST and calculation of the normalized difference vegetation index (NDVI) using the cloud computing-based platform (Google Earth Engine—GEE) (Velastegui-Montoya et al. 2023); (II) LULC Acquisition; and (III) Data statistical analysis was performed.

The following sections describe all phases. The codes used to retrieve LST and calculate NDVI on GEE are available for Landsat-8 OLI/TIRS (https://code.earthengine.google.com/70145f1cd2f3621eca03b928c1dcf47f) and Landsat-5 TM (https://code.earthengine.google.com/1f8a278eaade43a2c747a6ec0bff6e3d).

2.3 *Data Processing*

2.3.1 LST Retrieval and NDVI Calculation

Images from Landsat Collection 2 atmospherically corrected surface reflectance were used through GEE. The research was conducted in 1985, 1996 (Landsat-5 TM), and 2023 (Landsat-8 OLI/TIRS). Image processing included the application of scale factors for the optical and thermal bands, in addition to the removal of clouds and cloud shadows using masks based on the QA_PIXEL band. The thermal bands of Landsat-5 TM and Landsat-8 OLI/TIRS were used to estimate the LST, bands 6 and 10, respectively. Band 11 of Landsat-8 OLI/TIRS is not used because it has less calibration than Band 10 (Raiol et al. 2024; Waleed and Sajjad 2022).

The procedures adopted to estimate the LST were based on Waleed and Sajjad (2022) and Waleed et al. (2023), in which the Thermal bands are used to convert Digital Numbers (DN) into spectral radiance (L_λ) using Eqs. (1) and (2). The resulting top of atmospheric radiance (L_λ) is in watts/(m^2 × ster × μm) shown in Eq. (1):

$$L_\lambda = \left(\frac{L_{\text{MAX}} - L_{\text{MIN}}}{QCAL_{\text{MAX}} - QCAL_{\text{MIN}}}\right) \times (QCAL - QCAL_{\text{MIN}}) + L_{\text{MIN}} \qquad (1)$$

where L_{MAX} = maximum spectral radiances (15.600 for TM), L_{MIN} = minimum spectral radiances (1.238 for TM), $QCAL_{\text{MAX}}$ = maximum Digital Number (DN) value (255), $QCAL_{\text{MIN}}$ = minimum Digital Number (DN) value (1), and $QCAL$ = Digital Number value of band 6.

The values of L_{MAX}, L_{MIN}, $QCAL_{MAX}$, and $QCAL_{MIN}$ are obtained from the metadata file attached to each Landsat image. The Landsat-8 OLI thermal band, the top of atmospheric radiance (L_λ), is calculated using the following Eq. (2):

$$L_\lambda = M_L \times QCAL + A_L \tag{2}$$

where M_L = multiplicative rescaling factor for specific band (0.0003342), $QCAL$ = digital numbers of band 10, and A_L = additive rescaling factor for specific band (0.1).

The top-of-atmosphere (TOA) brightness temperature was calculated from spectral radiance using the following Eq. (3):

$$T_B = \frac{K_2}{\ln\left(\frac{K_1}{L_2} + 1\right)} \tag{3}$$

where T_B = at-satellite brightness temperature in Kelvin (K), L_λ = spectral radiance, K_1 and K_2 = calibrated constants depending on the sensor, and it can be found in the metadata files.

The emissivity correction depends on the category of land use, and it is evaluated using the Normalized Difference Vegetation Index (NDVI) per pixel value (Waleed and Sajjad 2022). Equation (4) is used to evaluate emissivity corrected LST:

$$S_T = \frac{T_B}{1 + \left(\frac{\lambda \times T_B}{\rho}\right) \times \ln\varepsilon} - 273.15 \tag{4}$$

where S_T = land surface temperature in (°C), T_B = at-satellite brightness temperature (K), λ = wavelength of emitted radiance (11.5 μm), ρ = 1.438 × 10 − 2 mK, and ε = emissivity (ranges from 0.97 to 0.99).

Emissivity can be calculated by using Eq. (5), which is

$$\varepsilon = 0.004Pv + 0.986 \tag{5}$$

where Pv = proportion of vegetation.

The proportion of vegetation Pv can be calculated using the following Eq. (6):

$$Pv = \left(\frac{\text{NDVI} - \text{NDVI}_{\min}}{\text{NDVI}_{\max} - \text{NDVI}_{\min}}\right)^2 \tag{6}$$

in which, NDVI, $\text{NDVI}_{\min}$, and $\text{NDVI}_{\max}$ are per pixel value of NDVI, minimum NDVI, and maximum NDVI values, respectively.

2.3.2 NDVI Calculation

NDVI was used to determine the spatiotemporal patterns of green areas in the study area (Chanuwan Wijesinghe et al. 2024). The range of NDVI values is − 1 to 1. Negative values in this contrast virtually represent the water bodies, while scores close to + 1 indicate dense vegetation.

NDVI can be measured using the reflectance values of the Visible (RED) and Near-Infrared (NIR) bands mentioned (Faisal et al. 2021), following Eq. (7):

$$NDVI = \frac{B_{NIR} - B_{RED}}{B_{NIR} + B_{RED}} \quad (7)$$

in which, NIR band represents Band 4 in Landsat TM [0.77–0.90 μm (wavelength) and Band 5 in Landsat OLI [0.85–0.88 μm (wavelength)] respectively, and the RED band represents Band 3 in Landsat TM [0.63–0.69 μm (wavelength)] and Band 4 in Landsat OLI [0.64–0.67 μm (wavelength)], respectively.

2.3.3 LULC Change

Land use and land cover (LULC) data were used from the MapBiomas platform in 1985, 1996, and 2023. This platform was acquired through GEE using collection 9.0, which provides LULC data for the entire Brazilian territory in raster format from 1985 to 2023 (MapBiomas 2024). Moreover, Class areas were quantified in hectares (ha) and percentages (%). The processing, systematization, and layout of maps were carried out by QGIS 3.34 software.

2.4 Statistical Analysis

The variability of LST was assessed for the years 1985, 1996, and 2023 across different LULC classes. For each LULC class and year, 25 random samples were collected, totaling 75 samples per LULC type. The LULC classes considered in the analysis included: Forest Formation, Floodable Forest, Pasture, Urban Area, Other Non-vegetated Areas, and Palm Oil Cultivation. These classes were selected because they represented the most significant areas (>1%) in the years studied. Subsequently, LST values were extracted for each LULC class.

Statistical analysis of the data was conducted in the R environment (R Core Team 2023). Initially, the *Shapiro–Wilk* test was applied to assess the normality of the LST data (considering a p-value < 0.05). Once the absence of normality and a non-monotonic relationship was confirmed, the *Kruskal–Wallis* test, a non-parametric method, was used to evaluate whether there were significant differences in LST means among at least one of the LULC classes. Finally, the post hoc*Dunn* test was performed to allow for multiple comparisons, identifying specific differences between the LULC classes. Spearman's correlation (ggplot2 package) verified the relationship between NDVI and LST in the years of analysis. This ranges from −1 to 1, with values close to 1 indicating a strong positive correlation between the variables (Raiol et al. 2025).

3 Results and Discussion

3.1 LULC Change Analysis

The spatial distribution of LULC changes (Fig. 2) shows a reduction in forest formation, mainly due to the expansion of pastures in the municipality of Capanema. In 1985 (Table 1), forest formation was predominant with 78.21% of the area, decreasing to 57.68% in 1996, and declining again, presenting 29.72% in 2023.

The pasture class represented 10.25% in 1985, subsequently showing a significant increase of 30.95% and 57.22% in 1996 and 2023, respectively. Another class that stood out was the urban area, presenting 1.21% in 1985 and 2.29% in 2023, representing an increase twice as large as the first year of analysis and reflecting an urban expansion. Over the years, a reduction in floodable forests and wetlands was also observed, which can be explained by the increase in the agricultural and pasture classes.

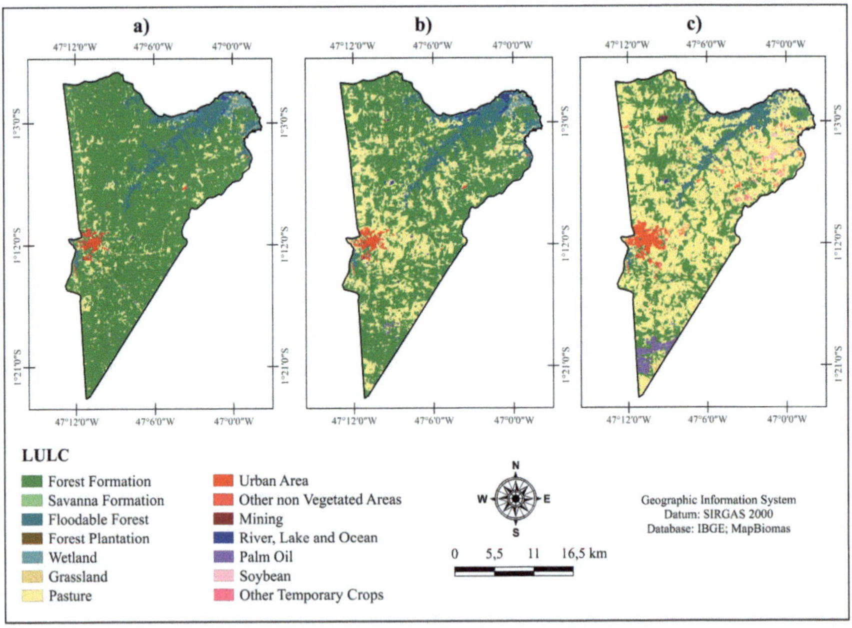

Fig. 2 LULC changes in the municipality of Capanema, **a** 1985, **b** 1996, and **c** 2023

Table 1 LULC quantification (% and ha) in the municipality of Capanema

Classes	1985		1996		2023	
	Area (km^2)	Area (%)	Area (km^2)	Area (%)	Area (km^2)	Area (%)
Forest formation	486.32	78.21	358.68	57.68	184.78	29.72
Savanna formation	0.01	0.002	0	0.00	0.01	0.002
Floodable forest	51.53	8.29	47.39	7.62	40.98	6.59
Forest plantation	0.00	0.00	0.00	0.00	0.02	0.00
Wetland	8.58	1.38	6.92	1.11	0.63	0.10
Grassland	2.22	0.36	2.13	0.34	0.22	0.03
Pasture	63.76	10.25	192.44	30.95	355.81	57.22
Urban area	7.55	1.21	8.62	1.39	14.23	2.29
Other non-vegetated areas	0.13	0.02	0.02	0.003	7.84	1.26
Mining	0.00	0.00	0.10	0.02	0.84	0.13
River, lake and ocean	1.43	0.23	4.56	0.73	1.22	0.20
Palm oil	0.27	0.04	0.98	0.16	10.47	1.68
Soybean	0.00	0.00	0.00	0.00	2.54	0.41
Other temporary crops	0.00	0.00	0.00	0.00	2.23	0.36

The expansion of pasture and agricultural areas has been the main driver of deforestation in several tropical regions, especially in the Amazon, where agriculture and livestock farming drive the conversion of natural ecosystems into productive areas (Fearnside 2006; Ferraz et al. 2023). This advance is largely due to the growing demand for agricultural and livestock commodities, driven by public policies that encourage the occupation of new areas and the expansion of agricultural frontiers (Rivero et al. 2009). The replacement of forests by pastures profoundly alters the dynamics of the environment, increasing LST due to reduced evapotranspiration and modification of the local energy balance, resulting in environmental problems that affect the resilience of ecosystems to environmental disturbances (Souza et al. 2022).

At the same time, the reduction of floodable and wetland forests, associated with the expansion of livestock farming and monocultures, has further exacerbated environmental impacts in the region (Santos et al. 2023). The inadequate occupation of these ecosystems contributes to the loss of essential habitats for several species, compromising the supply of fishing resources and increasing the vulnerability of communities that depend on these systems for their livelihoods (Soares et al. 2018). In addition, the degradation of floodplain forests intensifies extreme events, such as coastal erosion and the greater frequency of storm surges and floods, making these regions even more susceptible to climate change (Neves and Muehe 2008).

3.2 LST and NDVI Analysis

The LST over the years (Fig. 3) has shown an increase, spreading in all directions of the study area, but especially in the year 2023, it is possible to observe a significant increase in temperature over the urban area due to the densification and impermeability of the soil of the urban conglomerates. Another aspect is the reduction of green areas and the increase of agricultural and pasture areas, which consequently affects the increase of LST.

In 1985 (Table 2), the LST of the municipality of Capanema varied between 25.80 °C and 40.10 °C, with an average of 30.40 °C ± 2.51 °C. In 1996, there was a decrease in the minimum (24.40 °C) and maximum temperatures (39.30 °C); however, there was an increase in the mean temperature, which was 31.80 °C ± 2.37 °C. In 2023, there was a significant increase in temperature, with a minimum LST of 29.60 °C and a maximum of 44.80 °C. This increase was reflected in the rise of the mean LST, with 35.10 °C ± 3.06 °C.

The conversion of forests into pastures and urban areas in the Amazon has caused significant climate changes, including an increase in thermal amplitude by approximately 2 °C, a reduction in moisture retention, and changes in the regional hydrological balance (Antonucci et al. 2025). In addition, the replacement of forests by cities generates even more significant impacts, with a 62.55% increase in albedo, making anthropized environments more vulnerable to climate extremes, highlighting the

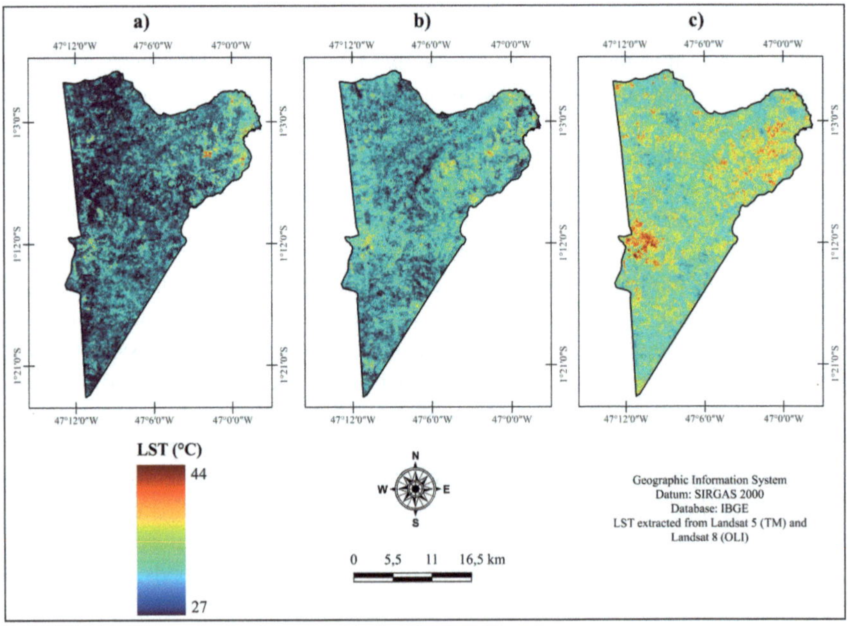

Fig. 3 LST variations in the municipality of Capanema, **a** 1985, **b** 1996, and **c** 2023

Table 2 LST Descriptive statistics for the municipality of Capanema

LST values	1985	1996	2023
Minimum temperature (°C)	25.80	24.40	29.60
Maximum Temperature (°C)	40.10	39.30	44.80
Mean (°C)	30.40	31.80	35.10
Standard deviation (°C)	2.51	2.37	3.06

importance of continuous monitoring to understand the effects of changes in LULC on the Amazon climate (Antonucci et al. 2025).

When analyzing the relative average increase in LST, the year 1996 compared to 1985 showed an average increase of 1.4 °C over 11 years. Between 1996 and 2023, this represented an increase of 3.30 °C in 27 years. And during the entire analysis period (1985–2023), it represented an increase of 4.70 °C, which corresponds to an annual increase of 0.12 °C over the last 38 years of analysis.

These findings are in line with other research developed in the west of the state of São Paulo (Brazil), where over 30 years of analysis, the LST increased 3.50 °C (0.117 °C per year) (Carrasco et al. 2020). Another result of increased LST was observed in Belém, the capital of the state of Pará (Brazil), where it increased by 1.5 °C in 37 years of analysis, representing an average rise of 0.04 °C per year (Raiol et al. 2024). This rate of increase of LST in the study area between 1985 and 2023 (4.70 °C) is above the global average temperature rate according to the Copernicus Climate Change Service (C3S), in which the year 2024 was 0.72 °C above the 1991–2020 average and 0.12 °C above 2023, the previous warmest year on record, which is equivalent to 1.60 °C above an estimate of the 1850–1900 temperature designated as the pre-industrial level (Copernicus 2025).

Urbanization significantly alters the natural environment, which has a direct impact on the local climate system, as it changes in urban structure, such as high population density and reduced vegetation, intensify temperature increases and promote the formation of heat islands, affecting the thermal comfort and well-being of the population (Rodrigues 2024). In addition, studies show that areas with less vegetation have higher temperatures, demonstrating the influence of urbanization in exacerbating the effects of global warming and creating regional climate hotspots (Rodrigues 2024).

The NDVI variation (Fig. 4) shows that there was an increase in values close to 0, that is, exposed areas or areas with little vegetation, mainly in the last year of analysis (2023), presenting a minimum value of 0.12, that is, lower than in 1985 (0.14) and 1996 (0.24). The maximum values varied between 0.85, 0.86, and 0.89 for 1985, 1996, and 2023, respectively. This increase in maximum values may be related to the expansion of pastures and agriculture. Thus, this may have also been reflected in the increase in average NDVI values, with 0.64 ± 0.15, 0.69 ± 0.12, and 0.69 ± 0.18 for 1985, 1996, and 2023, respectively. The greater variability in 2023 reflects the diversity of LULC, with urban, agricultural, and forest areas coexisting in the municipality of Capanema.

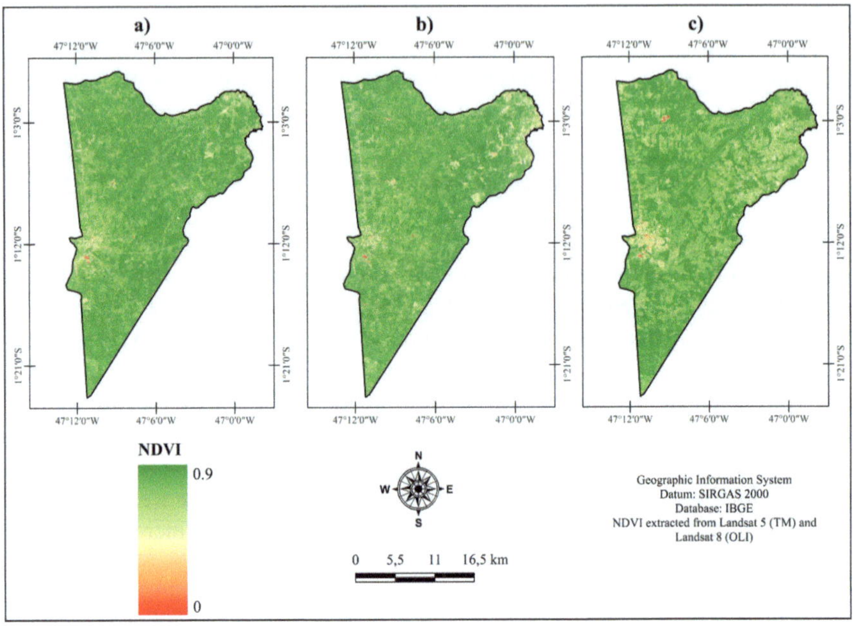

Fig. 4 NDVI variations in the municipality of Capanema, **a** 1985, **b** 1996, and **c** 2023

Another relationship observed was the Spearman correlation between NDVI and LST (Fig. 5), in which the year 1985 was significant (p = −0.37; *p-value* < 0.05), but considered moderate to 1996 (p = −0.45; *p-value* < 0.05), and 2023 (p = −0.82; *p-value* < 0.05). This relationship between LST and NDVI can be highlighted in 1985, when most of the area was covered by forests (78.21%), which had high NDVI and low LST. However, the presence of other classes (such as pastures and urban areas) with lower NDVI and higher LST may have diluted the correlation, resulting in a moderate negative relationship.

The reduction of forests to 57.68% in 1996 and the increase of grasslands to 30.95% created a greater contrast between areas with high NDVI (forests) and low NDVI (pasture and urban areas), intensifying the negative relationship between NDVI and LST. By 2023, the drastic loss of forests (only 29.72%) and the dominance of grasslands (57.22%) created a scenario where areas with high NDVI (remaining forests) have low LST, and areas with low NDVI (pasture, urban areas, and agricultural areas) have high LST. This polarization increased the strength of the negative correlation.

This negative correlation between LST and NDVI has been observed in several studies, where factors such as LULC, bare soil, vegetation loss, and urban areas influence higher LST values and carbon stocks (Faisal et al. 2021; Kafy et al. 2022; Rahaman et al. 2022). Vegetation plays a critical role in moderating LST and UHI, especially in urban and semi-urban areas, and a reduction in intensive human activities affects emissions of pollutants to the atmosphere from automobiles and industry

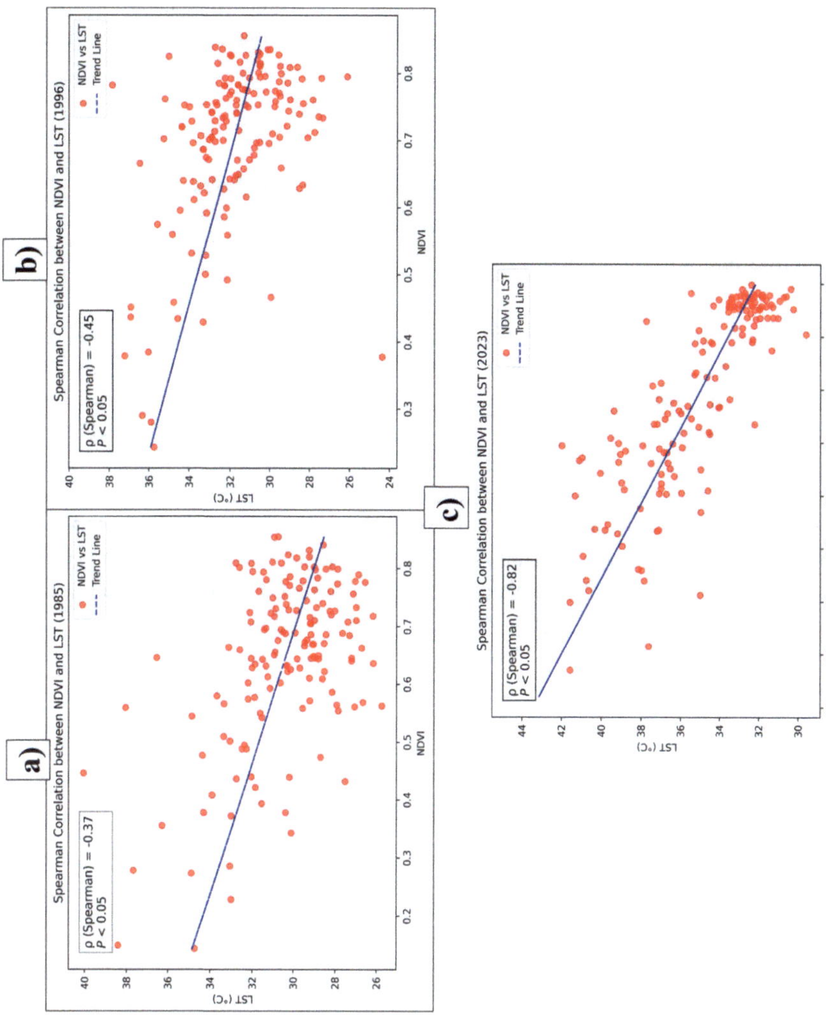

Fig. 5 Spearman's Correlation between LST and NDVI in the municipality of Capanema, **a** 1985, **b** 1996, and **c** 2023

(Veettil et al. 2023). It reduces greenhouse gases and stabilizes thermal conditions and vegetation health in cities (Veettil et al. 2023). Understanding these dynamics is essential for urban planning and climate adaptation strategies.

Thus, it can be seen that LST and NDVI are indicators that make it possible to visualize environmental degradation and, consequently, the loss of ecological resilience, reflecting not only the conversion of vegetation but also the reduction in the capacity of the soil and vegetation cover to regulate thermal extremes, compromising the stability of ecosystems and the mitigation of local climate variations (Oliveira et al. 2025).

3.3 *Relationship Between LST and LULC: Analysis of Differences Between LST in Response to LULC Classes*

The LST variation in different types of LULC (Fig. 6a) and in the years of analysis presented significant differences (Fig. 6b). These significant differences were confirmed by the Kruskal–Wallis test, being between the LST groups for different classes (*chi-squared* = 135.07, *df* = 5, *p-value* < 0.05) and also for the LST groups for the years of analysis (*chi-squared* = 166.65, *df* = 2, *p-value* < 0.05).

Dunn's test indicated a significant difference between the classes: Other non-vegetated Areas—Forest Formation; Other non-vegetated Areas—Floodable Forest; Pasture—Forest Formation; Pasture—Floodable Forest; Pasture—Palm Oil; Urban Area—Forest Formation; Urban Area—Floodable Forest; Urban Area—Other non-vegetated Areas; and Urban Area—Palm Oil (*p-adj* < 0.05). Therefore, it is possible to observe that the classes Other non-vegetated Areas, Pasture, and Urban Area were the ones that stood out the most, also having higher mean LST values with 33.10 °C ± 3.43 °C, 33.70 °C ± 2.99 °C, and 35.50 °C ± 3.44 °C, respectively. Other studies

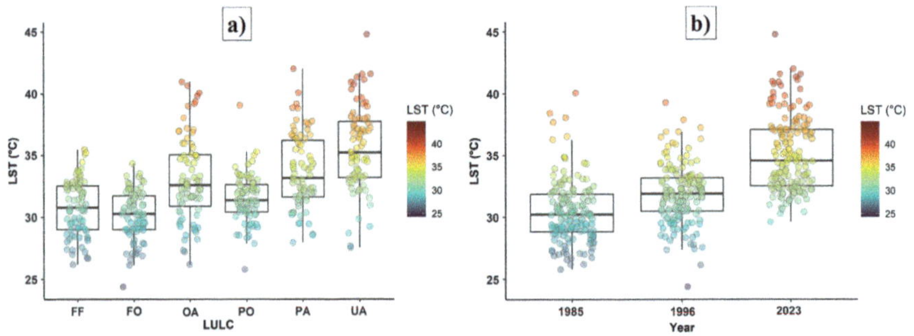

Fig. 6 **A** Boxplot graph of the LST average for different LULC classes, **b** Boxplot graph of the LST mean in 1985, 1996, and 2023. Abbreviations: Forest Formation (FO), Floodable Forest (FO), Other non-vegetated Areas (OA), Palm Oil (PO), Pasture (PA), and Urban Area (UA)

have also found high values for urban areas, agricultural areas, and pastures ranging from 32 °C to 37 °C, which is close to the results found in this research (Balew and Korme 2020; Faisal et al. 2021; M. P. Ribeiro et al. 2024).

In the analysis of groups considering the transition years 1985–1996, 1996–2023, and 1985–203, there were significant differences between all groups (*p-adj* < 0.05), the most considerable being between 1985 and 2023, which corresponds to an increase of 4.70 °C (increase of 0.12 °C per year) over the 38 years of analysis. Statistical tests between comparison of groups considering years of analysis also observed significant differences in several regions of the world, with a trend to increase over the years, which is consistent with the Sixth Assessment Report (AR6) of the Intergovernmental Panel on Climate Change (IPCC) (IPCC 2023; Raiol et al. 2024; Waleed and Sajjad 2022).

These results reveal a clear trend of increasing LST over the past decades, likely associated with changes in LULC, such as urbanization and deforestation, alongside global climate factors like global warming. The most pronounced increase observed between 1985 and 2023 underscores the urgent need for mitigation policies to address the impacts of climate change and unsustainable land use practices. In light of this scenario, the implementation of effective mitigation and adaptation strategies is crucial to minimize the adverse effects of rising temperatures and to enhance environmental resilience in both urban and rural areas.

In urban areas, strategies such as expanding vegetation cover, modifying urban geometry, and using materials that reduce heat retention demonstrate significant potential for mitigating the effects of excessive heat (Ferrarezi et al. 2023). In rural contexts, the adoption of sustainable agricultural practices, such as integrated crop–livestock–forestry systems and the restoration of degraded pastures, is essential to reduce greenhouse gas emissions and enhance climate resilience (Assad and Assad 2024). These initiatives should be reinforced by public policies for land use planning and incentives for low-impact agricultural production, aiming to promote sustainable development in the Eastern Amazon.

4 Conclusions

This study analyzed Land Surface Temperature (LST) and its relationship with changes in Land Use and Land Cover (LULC) in the municipality of Capanema, Eastern Amazon, using Landsat thermal sensors. The results confirmed the influence of LULC changes on LST variation, revealing a significant upward trend in temperature over the past decades. This increase was particularly pronounced between 1985 and 2023, during which the average temperature rose by 4.7 °C, reflecting the impact of the conversion of forested areas into pastures and urban zones. The expansion of pastures and urban areas, combined with the reduction of natural vegetation, intensified the heat island effect and contributed to the rise in LST.

The results indicated a significant negative correlation between NDVI and LST, demonstrating that the reduction in vegetation cover is directly associated with an increase in temperature. Furthermore, statistical analysis confirmed significant differences in LST among different LULC classes, with urban areas and pastures exhibiting the highest temperatures. These findings underscore the need for mitigation and adaptation strategies, such as sustainable urban planning and the restoration of degraded areas, to minimize the impacts of rising temperatures in the region. Additionally, they highlight the importance of remote sensing and statistical analyses for environmental monitoring, providing valuable insights for public policies aimed at conservation and sustainable development in the Eastern Amazon.

Declaration of Competing Interest The authors declare that they have no known competing financial interests or personal relationships that could have appeared to influence the work reported in this paper.

References

Alvares CA, Stape JL, Sentelhas PC, de Moraes Gonalves JL, Sparovek G (2013) Köppen's climate classification map for Brazil. Meteorol Z 22(6):711–728. https://doi.org/10.1127/0941-2948/2013/0507

Antonucci B, de Neves GZF, de Andrade NLR, Aguiar RG, Webler AD (2025) Microclima no Cenário de Floresta, Pastagem e Cidade no Sudoeste da Amazônia. Revista Brasileira de Meteorologia 40. https://doi.org/10.1590/0102-778640230038

Assad ED, Assad MLRCL (2024) Mudanças do clima e agropecuária: impactos, mitigação e adaptação. Desafios e oportunidades. Estudos Avançados 38(112):271–292. https://doi.org/10.1590/s0103-4014.202438112.015

Balew A, Korme T (2020) Monitoring land surface temperature in Bahir Dar city and its surrounding using Landsat images. Egypt J Remote Sens Space Sci 23(3):371–386. https://doi.org/10.1016/j.ejrs.2020.02.001

Bokaie M, Zarkesh MK, Arasteh PD, Hosseini A (2016) Assessment of urban heat Island based on the relationship between land surface temperature and land use/land cover in Tehran. Sustain Cities Soc 23:94–104. https://doi.org/10.1016/j.scs.2016.03.009

Carrasco RA, Pinheiro MMF, Junior JM, Cicerelli RE, Silva PA, Osco LP, Ramos APM (2020) Land use/land cover change dynamics and their effects on land surface temperature in the western region of the state of São Paulo, Brazil. Reg Environ Chang 20(3):96. https://doi.org/10.1007/s10113-020-01664-z

Chanuwan Wijesinghe D, Withanage NC, Mishra PK, Frenando WS, Abdelrahman K, Fnais MS (2024) The impacts of landscape structure changes on urban surface temperature and heat islands formation of a growing city in southern Sri Lanka. Front Environ Sci 12. https://doi.org/10.3389/fenvs.2024.1474742

Copernicus (2025) Copernicus—Climate change service. Copernicus: 2024 is the first year to exceed 1.5 °C above pre-industrial levels. https://climate.copernicus.eu/copernicus-2024-first-year-exceed-15degc-above-pre-industrial-level#9af2bdf9-8c2d-4baa-bc6b-02d8572ab46c. Last accessed 11 Feb 2025

de Almeida CR, Teodoro AC, Gonçalves A (2021) Study of the urban heat island (UHI) using remote sensing data/techniques: a systematic review. Environments 8(10):105. https://doi.org/10.3390/environments8100105

Souza de VAS, Rotunno Filho OC, Rodriguez DA, Moreira DM, Rudke AP, Andrade CD (2022) Dinâmica da conversão de floresta e tendências climáticas na bacia do rio Madeira. Ciência Florestal 32(4):2007–2034. https://doi.org/10.5902/1980509865211

Diem PK, Nguyen CT, Diem NK, Diep NTH, Thao PTB, Hong TG, Phan TN (2024) Remote sensing for urban heat island research: Progress, current issues, and perspectives. Remote Sens Appl: Soc Environ 33:101081. https://doi.org/10.1016/j.rsase.2023.101081

Duan S-B, Li Z-L, Leng P (2017) A framework for the retrieval of all-weather land surface temperature at a high spatial resolution from polar-orbiting thermal infrared and passive microwave data. Remote Sens Environ 195:107–117. https://doi.org/10.1016/j.rse.2017.04.008

Espinoza J-C, Jimenez JC, Marengo JA, Schongart J, Ronchail J, Lavado-Casimiro W, Ribeiro JVM (2024) The new record of drought and warmth in the Amazon in 2023 related to regional and global climatic features. Sci Rep 14(1):8107. https://doi.org/10.1038/s41598-024-58782-5

Faisal A-A-, Kafy A-A, Al Rakib A, Akter KS, Jahir DMA, Sikdar MS et al (2021) Assessing and predicting land use/land cover, land surface temperature and urban thermal field variance index using Landsat imagery for Dhaka Metropolitan area. Environ Chall 4:100192. https://doi.org/10.1016/j.envc.2021.100192

Fapespa (2024) Fundação Amazônia de Amparo a Estudos e Pesquisas. Anuário estatístico do Pará 2025. Estatística Municipal. Capanema. Fapespa. Estatística Municipal: Capanema. https://www.fapespa.pa.gov.br/wp-content/uploads/2024/04/Capanema.pdf. Last accessed 1 Feb 2025

Fearnside PM (2006) Desmatamento na Amazônia: dinâmica, impactos e controle. Acta Amazon 36(3):395–400. https://doi.org/10.1590/S0044-59672006000300018

Ferrarezi de HS, Sousa de MNPO, Monteiro EZ (2023) Estratégias de mitigação de calor em climas tropicais. VERNÁCULA—Territórios Contemporâneos 1(1). https://doi.org/10.18312/verncula.v1i1.2306

Ferraz AMM, Pires ÉG, Barbosa e Souza L (2023) Alterações na temperatura de superfície em áreas de expansão do agronegócio sobre um território quilombola no Tocantins. Interações (Campo Grande), pp 975–989. https://doi.org/10.20435/inter.v24i3.3795

Fleischmann A, Papa F, Hamilton S, Melack J, Forsberg B, Val A et al (2024) Extreme warming of Amazon waters in a changing climate. Earth. https://doi.org/10.31223/X56D9T

Furtado LS, Pereira RVS, de Souza EB (2024) Hemeroby mapping of the Belém landscape in eastern Amazon and impact study of urbanization on the local climate. Urban Science 8(1):15. https://doi.org/10.3390/urbansci8010015

IBGE (2022) IBGE Cidades. Instituto Brasileiro de Geografia e Estatística. https://cidades.ibge.gov.br/brasil/pa/novo-progresso/panorama. Last accessed 19 Jun 2024

IPCC (2023) Sections. In: Lee H, Romero J (eds) [Core Writing Team] Climate change 2023: synthesis report. Contribution of working groups i, ii and iii to the sixth assessment report of the intergovernmental panel on climate change. Geneva, Switzerland. https://www.ipcc.ch/report/ar6/syr/. Last accessed 11 Feb 2025

Kafy A-Al, Faisal A-A-, Al Rakib A, Md Fattah A, Rahaman ZA, Sattar GS (2022) Impact of vegetation cover loss on surface temperature and carbon emission in a fastest-growing city, Cumilla, Bangladesh. Build Environ 208:108573. https://doi.org/10.1016/j.buildenv.2021.108573

Malakar NK, Hulley GC, Hook SJ, Laraby K, Cook M, Schott JR (2018) An operational land surface temperature product for landsat thermal data: methodology and validation. IEEE Trans Geosci Remote Sens 56(10):5717–5735. https://doi.org/10.1109/TGRS.2018.2824828

MapBiomas (2024) Plataforma MapBiomas v.8.0. https://brasil.mapbiomas.org/. Accessed 19 June 2024

Marengo JA, Cunha AP, Espinoza J-C, Fu R, Schöngart J, Jimenez JC et al (2024) The drought of Amazonia in 2023–2024. Am J Clim Chang 13(03):567–597. https://doi.org/10.4236/ajcc.2024.133026

Moreira DM, Papa F, Fassoni-Andrade A, Fleischmann AS, Wongchuig S, Paiva RCD de et al (2024) Widespread and exceptional reduction in river water levels across the Amazon Basin during the 2023 extreme drought revealed by satellite altimetry and SWOT. ESS Open Archive. https://doi.org/10.22541/essoar.172408211.17914763/v1

Naughton J, McDonald W (2019) Evaluating the variability of urban land surface temperatures using drone observations. Remote Sens 11(14):1722. https://doi.org/10.3390/rs11141722

Neves C, Muehe D (2008) Vulnerabilidade, impactos e adaptação às mudanças do clima: a zona costeira. In: Pires T (ed) Parcerias Estratégicas: Mudança do clima no Brasil: vulnerabilidade, impactos e adaptação. Brasília: Parcerias Estratégicas, pp 217–295

Oliveira JS, Ribeiro AA, Gamarra RM, Paranhos Filho AC (2025) Relação entre vegetação urbana e temperatura de superfície através de uma correlação NDVI—LST. Rev Políticas Públicas Cid 14(1):e1291. https://doi.org/10.23900/2359-1552v14n1-28-2025

Rahaman ZA, Kafy A-A, Saha M, Rahim AA, Almulhim AI, Rahaman SN et al (2022) Assessing the impacts of vegetation cover loss on surface temperature, urban heat island and carbon emission in Penang city, Malaysia. Build Environ 222:109335. https://doi.org/10.1016/j.buildenv.2022.109335

Raiol LL, Santos DCR, Santos Silva KW, Silva LP, Ferreira MT, Gomes JA et al (2019) Gerenciamento de pneus inservíveis na cidade de Capanema, Pará. Braz J Dev 5(12):30989–31004. https://doi.org/10.34117/bjdv5n12-204

Raiol LL, da Rocha YAS, Santos DCR, de Lima AMM, Velastegui-Montoya A (2024) Assessment of urban heat islands in an eastern Amazonian city. The international archives of the photogrammetry, remote sensing and spatial information sciences XLVIII-3-2024:437–443. https://doi.org/10.5194/isprs-archives-XLVIII-3-2024-437-2024

Raiol LL, da Rocha YAS., de Lima AMM, Velastegui-Montoya A (2025) Assessment of environmental degradation and conservation in the Maracanã River Basin, eastern amazon. Front Environ Sci 12. https://doi.org/10.3389/fenvs.2024.1506701

R Core Team (2023) R: a language and environment for statistical computing. R Foundation for Statistical Computing, Vienna, Austria. (Available at: https://www.R-project.org/). Accessed 5 June 2024

Ribeiro MP, Menezes GP, Figueiredo GKDA, de Mello K, Valente RA (2024) Impacts of urban landscape pattern changes on land surface temperature in Southeast Brazil. Remote Sens Appl: Soc Environ 33:101142. https://doi.org/10.1016/j.rsase.2024.101142

Ribeiro WO (2018) Cidade de porte médio de responsabilidade territorial: particularidades de Capanema no Nordeste do Pará/middle level city with territorial responsibility: particularities of Capanema in Northeast Pará. Revista GeoAmazônia 5(10):110. https://doi.org/10.18542/geo.v5i10.12503

Rivero S, Almeida O, Ávila S, Oliveira W (2009) Pecuária e desmatamento: uma análise das principais causas diretas do desmatamento na Amazônia. Nova Economia 19(1):41–66. https://doi.org/10.1590/S0103-63512009000100003

Rodrigues JEC (2024) Análise espacial da temperatura da área central da cidade de Belém (PA). Revista Brasileira de Climatologia 35:529–546. https://doi.org/10.55761/abclima.v35i20.18699

Santos MC, Do Canto O, Sombra D, Da Silva CN, Gusmão LHA, Fenzl N, Bastos RZ (2023) Ocupações de manguezais: risco, vulnerabilidade e território usado na zona costeira paraense. Rev Gestão E Secr (Manag Adm Prof Rev) 14(8):12554–12572. https://doi.org/10.7769/gesec.v14i8.2554

Soares D, Castro C, Mendes R, Moraes S, Pinho D, Mergulhão L (2018) Território usado e recursos hídricos na Amazônia brasileira os múltiplos usos do território na bacia hidrográfica do rio Apeú (Pará/Brasil). Revista Caribeña de Ciencias Sociales (RCCS), pp 1–24

Veettil BK, Puri V, Van DD, Quang NX (2023) Variations in land surface temperatures in small-scale urban areas in Vietnam during Covid-19 restrictions: case studies from Da Nang, Hue and Vinh City. Environ Monit Assess 195(7):822. https://doi.org/10.1007/s10661-023-11332-4

Velastegui-Montoya A, Montalván-Burbano N, Carrión-Mero P, Rivera-Torres H, Sadeck L, Adami M (2023) Google earth engine: A global analysis and future trends. Remote Sens 15(14):3675. https://doi.org/10.3390/rs15143675

Waleed M, Sajjad M (2022) Leveraging cloud-based computing and spatial modeling approaches for land surface temperature disparities in response to land cover change: evidence from Pakistan. Remote Sens Appl: Soc Environ 25:100665. https://doi.org/10.1016/j.rsase.2021.100665

Waleed M, Sajjad M, Acheampong AO, Alam MdT (2023) Towards sustainable and livable cities: leveraging remote sensing, machine learning, and geo-information modelling to explore and predict thermal field variance in response to urban growth. Sustainability 15(2):1416. https://doi.org/10.3390/su15021416

Wang M, Zhang Z, Hu T, Wang G, He G, Zhang Z et al (2020) An efficient framework for producing landsat-based land surface temperature data using google earth engine. IEEE J Sel Top Appl Earth Obs Remote Sens 13:4689–4701. https://doi.org/10.1109/JSTARS.2020.3014586

Yang H, Xi C, Zhao X, Mao P, Wang Z, Shi Y et al (2020) Measuring the urban land surface temperature variations under Zhengzhou City expansion using landsat-like data. Remote Sens 12(5):801. https://doi.org/10.3390/rs12050801

Ye X, Zhu J, Zhu J, Duan Y, Wang P (2024) Comparison of nighttime land surface temperature retrieval using mid-infrared and thermal infrared remote sensing data under different atmospheric water vapor conditions. IEEE Trans Geosci Remote Sens 62:1–9. https://doi.org/10.1109/TGRS.2024.3399010

Zhang D, Tang R, Zhao W, Tang B, Wu H, Shao K, Li Z-L (2014) Surface soil water content estimation from thermal remote sensing based on the temporal variation of land surface temperature. Remote Sens 6(4):3170–3187. https://doi.org/10.3390/rs6043170

Land Use Land Cover Analysis: Understanding the Relationship Between Urban Development and Urban Heat Islands

V. Anitha and D. Prabha

Abstract Urbanization is a dominant anthropogenic process reshaping Earth's natural system, significantly influencing Land Use and Land Cover (LULC) and contributing to the Urban Heat Island (UHI) effect. The UHI effect refers to the phenomenon where urban areas are warmer than their surrounding rural or suburban regions due to replacing natural landscapes with impervious surfaces. This study emphasizes the role of LULC changes and their implications for surface temperature dynamics. Rapid urban growth, particularly in developing regions such as Guyana, India, Argentina, and Ethiopia, has led to the expansion of impervious surfaces, replacing vegetative and permeable land, thereby intensifying UHI effects. Advances in geospatial tools, including supervised and unsupervised classification techniques combined with thermal remote sensing, enable precise monitoring and mapping of LULC transformations. The findings emphasize the necessity of integrating long-term, high-resolution data to understand urban climate impacts better, guide sustainable urban planning, and mitigate UHI effects.

Keywords Urban Heat Islands · Land use changes · Urbanization · Impervious surfaces · Urban sprawl

1 Introduction

Urbanization is one of the most dominating anthropogenic changes, exerting extensive influence on the Earth's natural system. Since the mid-1980s, rapid urban growth has been observed in many cities across developing countries, which are currently experiencing the fastest rates of urban expansion. The dramatic increase in urban land has led to significant changes in Land Use and Land Cover (LULC) patterns

V. Anitha (✉) · D. Prabha
Department of Environmental Sciences, Bharathiar University, Coimbatore, India
e-mail: anithavijayakumar8@gmail.com

D. Prabha
e-mail: prabha.ens@buc.edu.in

B. Kozhikkodan Veettil and B. Das (eds.), *Remote Sensing of Urban Heat Islands*, Cities and Nature, https://doi.org/10.1007/978-3-032-11580-5_5

(Tesfamariam et al. 2023). These LULC Changes have been directly linked to rising Land Surface Temperature (LST) and are considered major contributors to the formation of the Urban Heat Island (UHI) effect. UHI refers to the phenomenon where urban areas are significantly warmer than their surrounding rural regions (Schwarz et al. 2011), leading to adverse environmental consequences (Wemegah et al. 2020), and reflecting the degradation of natural ecosystems. This occurs as cities replace natural land covers, such as vegetation, water bodies and wetlands, with dense infrastructures like buildings, pavement, and other heat-absorbing and retaining surfaces.

The UHI effect has been extensively studied in various cities worldwide. With advancements in satellite technology, surface temperature data can now be retrieved at different temporal and spatial scales over both land and water surfaces. The Moderate Resolution Imaging Spectroradiometer (MODIS) satellite has been continuously providing data since 2000, offering a spatial resolution of 1 km and a temporal resolution of 2–3 days. Additionally, the Landsat satellite missions, operated by NASA and the USGS, have provided valuable data for studying LULC changes and UHI since the launch of Landsat 1 in 1972. Sensors such as the Thematic Mapper (TM), Enhanced Thematic Mapper Plus (ETM+), and Operational Land Imager (OLI) offer higher spatial resolution (30 m for multispectral bands and 60–120 m for thermal bands), making them widely used for monitoring urban development, LULC changes, and LST. The integration of MODIS and Landsat data enables researchers to analyze both broad temporal trends and fine-scale spatial variations in urban environments. Key parameters such as surface temperature, the Normalized Difference Vegetation Index (NDVI), and LULC maps are commonly retrieved using satellite data.

Surface temperature, the Normalized Difference Vegetation Index (NDVI), the Normalized Difference Water Index (NDWI), various built-up indices, and Land Use/Land Cover (LULC) maps are commonly derived from MODIS and Landsat data. Additionally, Sentinel-2 satellites, operated by the European Space Agency (ESA), provide high-resolution multispectral imagery, which is particularly useful for identifying LULC classes.

Several studies have explored the relationship between surface temperature and LULC changes in different urban areas. However, limited research has been conducted to investigate the relationship between UHI and LULC changes in smaller cities or rapidly urbanizing regions, particularly in developing countries. The integration of MODIS, Landsat, and Sentinel data allows for a more comprehensive understanding of these dynamics, highlights the importance of long-term, high-resolution satellite monitoring for urban climate studies.

Land Use and Land Cover Dynamics

Land cover is referred to as the spatial distribution of soil, water, flora, and on the earth's surface, whereas land use refers to how humans manage or modify the land (Lambin and Geist 2008). Land Use and Land Cover Change (LULCC) is a visible outcome of urbanization, profoundly impacting the environment and human well-being. It involves transformations such as deforestation, soil degradation, and agricultural shifts, which reduce biological productivity and environmental sustainability (Sayão et al. 2020). These changes contribute significantly to climate change

(Yang et al. 2013), biodiversity loss, and health risks (Melese 2016; Bellard et al. 2012). Urbanization and industrialization often replace green vegetation with impervious surfaces (Mahato and Pal 2018), altering the landscape and hydrologic cycles by increasing surface runoff, depleting groundwater (Dutta and Guchhait 2022), and raising Land Surface Temperature (LST) (Imran et al. 2021). The transition to built-up areas modifies surface properties like albedo and emissivity, intensifying the Urban Heat Island (UHI) effect (Spångmyr 2010). Monitoring LULCCs is essential for understanding environmental impacts, managing natural resources, and planning sustainable development. Advancements in remote sensing and geospatial technologies have enhanced the ability to detect and analyze these changes over time.

Key LULC Classes in Urban Areas

- **Built-up areas**: Characterized by impervious surfaces such as concrete, asphalt, and buildings, these areas are primary contributors to UHI formation due to their low albedo and high heat storage capacity.
- **Green spaces**: Forests, gardens, and parks within urban areas that provide cooling effects through evapotranspiration and shading, thereby mitigating UHI intensity.
- **Water bodies**: Rivers, lakes, and reservoirs that influence local microclimates by moderating temperature fluctuations.
- **Bare soil**: Exposed soil surfaces that exhibit intermediate thermal properties, influencing UHI effects depending on their extent and composition.

Urbanization Trends and LULC Transformations

Rapid urbanization often leads to the expansion of built-up areas at the expense of vegetative and permeable surfaces. This transformation exacerbates UHIs by increasing surface temperatures and altering local energy balances. Studies have documented that urban areas with higher impervious surface fractions exhibit stronger UHI effects (Peng et al. 2012).

Factors Driving Urban Development

Urbanization is driven by multiple factors, including population growth, through migration for better job opportunities and living standards, as well as natural population increase. Economic drivers such as industrialization, globalization, and growth in the service sector encourage urban expansion. Infrastructure improvements, including transportation networks, utilities, and housing projects, also play a critical role. Political and administrative policies, such as urban development initiatives and foreign investments, contribute to growth. Technological advancements, including smart cities and innovation hubs, further fuel development. Social changes, like the demand for better lifestyles and access to education, are also significant drivers. Environmental factors such as strategic location and resource availability can enhance urbanization, while tourism and the preservation of cultural heritage sites attract investment. Global trends, including urban sprawl and redevelopment also influence development patterns, alongside the rising demand for housing and commercial spaces.

Relationship between LULC and UHIs

The relationship between Land Use and Land Cover (LULC) changes and Urban Heat Islands (UHIs) is a crucial aspect of urban climate studies. The transformation of natural landscapes into built-up areas significantly alters surface energy balance, temperature distribution, and local climatic conditions.

LULC Changes and Their Influence on UHIs

Urban expansion leads to changes in land cover, replacing vegetative surfaces with impervious materials like asphalt and concrete. These materials have high thermal inertia and low albedo, causing them to absorb and retain heat during the day and slowly release it at night. Consequently, cities experience higher temperatures compared to their surrounding rural areas, forming UHIs.

Role of Vegetation Loss in UHI Formation

Vegetation plays a vital role in regulating urban temperatures through evapotranspiration and shading. The conversion of green spaces into commercial or residential zones reduces these cooling effects, exacerbating the UHI phenomenon. Studies have shown that cities with lower vegetation cover exhibit higher Land Surface Temperatures, especially during summer.

Impervious Surfaces and Heat Retention

As urbanization progresses, the proportion of impervious surfaces increases. These surfaces restrict natural water infiltration and limit soil moisture availability, further reducing evaporative cooling. The high heat retention capacity of these materials amplifies nighttime temperatures, prolonging heat stress conditions in urban environments.

Land Cover Transitions ad Spatial UHI Variations

Different types of LULC transitions contribute to spatial variations in UHI intensity. For instance, the conversion of agricultural land into industrial zones often results in higher temperature surges compared to residential developments with green infrastructure. Additionally, high-density urban areas with limited open spaces tend to exhibit stronger UHIs than low-density, mixed-use developments.

Mitigation Strategies Through LULC Management

Understanding the relationship between LULC and UHIs is essential for sustainable urban planning. Key mitigation strategies include:

- **Increasing urban vegetation**: Integrating green infrastructure and increasing tree cover can enhance cooling.
- **Adopting reflective roofing materials**: High-albedo surfaces reduce heat absorption.
- **Smart land use planning**: Policies promoting urban forest and permeable pavements can help reduce heat accumulation.

Case Studies and Empirical Evidence

Several studies have demonstrated the impact of LULC on UHIs. For instance, research in tropical and temperate cities has revealed a direct correlation between vegetation reduction and temperature rise in Kottayam (Anitha et al. 2023) and Alappuzha (Anitha et al. 2024). Similarly, studies have shown a link between built-up expansion and temperature rise in Delhi (Pramanik and Punia 2020) and Mumbai (Kavathekar et al. 2024). Comparative analysis of cities with varied LULC patterns highlights how strategic urban planning can influence urban thermal environments.

Surface Energy Balance

LULC changes alter the surface energy balance, influencing the partitioning of energy into sensible and latent heat. Impervious surfaces enhance sensible heat fluxes, while vegetative areas promote latent heat fluxes through evapotranspiration.

Spatial Patterns of UHIs

The intensity and spatial distribution of UHIs are closely linked to LULC patterns. Dense urban cores with high building densities exhibit the highest temperatures, while peripheral green spaces and water bodies act as cooling zones. Studies have shown a positive correlation between impervious surface area and Land Surface Temperature (Zhang et al. 2019).

Temporal Dynamics

Seasonal variations influence the UHI effect, with stronger intensities observed during summer due to higher solar radiation. Diurnal variations are also significant, with nighttime UHIs often more pronounced due to reduced radiative cooling in urban areas.

Urban Development and its Impact on UHIs

Urban development transforms natural landscapes into impervious surfaces like roads, buildings, and parking lots. Several mechanisms contribute to the formation of UHIs:

1. **Surface Characteristics**: Impervious materials have high thermal conductivity and low albedo, absorbing more solar radiation and retaining heat.
2. **Anthropogenic Heat**: Energy consumption for heating, cooling, transporting, and industrial activities contributes to heat generation.
3. **Reduced Vegetation**: Urbanization often leads to deforestation and loss of green spaces, decreasing evaporative cooling and carbon sequestration.
4. **Geometry of Urban Areas**: The "urban canyon" effect, created by densely packed high-rise buildings, traps heat and reduces wind flow.

Quantifying UHI Intensity

UHI intensity is typically measured as the temperature difference between urban and rural areas. Remote sensing-derived LST data, coupled with meteorological measurements, provide insights into spatial and temporal variations of UHIs.

Mitigating Urban Heat Islands Through LULC Management

Urban Heat Islands (UHIs) are a significant environmental and public health concern, characterized by elevated temperatures within urban areas compared to surrounding rural areas. Land Use and Land Cover (LULC) management strategies are crucial in mitigating these effects. Here's an expanded look at key approaches:

1. **Increasing Vegetation Cover**

Tree Planting: Strategic planting of trees, particularly deciduous trees that provide shade in summer and allow sunlight in winter, can significantly reduce surface temperatures.

Urban Forests: Establishing urban forests and green belts not only provides cooling but also improves air quality by absorbing pollutants.

Green Roofs and Walls: These features increase shading, enhance evapotranspiration (the process of water evaporating from plants and soil), and reduce heat absorption by buildings.

2. **Creation of Green Spaces**

Parks and Gardens: These provide vital cooling spaces for residents, reducing heat stress and improving overall well-being.

Water Bodies: Lakes, ponds, and even small water features contribute to cooling through evaporation and provide a visual cooling effect.

Urban Agriculture: Integrating urban agriculture into the cityscape can increase vegetation cover and provide a source of fresh produce.

3. **Establishing Cool Surfaces**

Cool Roofs: Using high-albedo materials (materials with high reflectivity) for roofs significantly reduces heat absorption and increases heat reflection, lowering building temperatures.

Cool Pavements: Implementing permeable pavements and using light-colored concrete or asphalt can reduce surface temperatures of roads and sidewalks.

4. **Sustainable Urban Planning**

Minimizing Impervious Surfaces: Reducing the area covered by asphalt, concrete, and other impermeable surfaces is critical. These surfaces absorb and retain heat, contributing significantly to UHI effects.

Maximizing Green Infrastructure: Incorporating green infrastructure elements such as bioswales, rain gardens, and green roofs into urban planning helps to manage stormwater runoff, reduce heat, and improve air quality.

Compact Development: Promoting compact development with mixed-use zoning encourages walkability and reduces the need for car travel, thus decreasing vehicle emissions and associated heat generation.

5. **Enhancing Urban Forestry Management**

Tree Selection and Placement: Careful selection of tree species considering their size, shade provision, and water needs is essential. Proper tree placement maximizes cooling benefits while minimizing conflicts with infrastructure.

Tree Maintenance: Regular tree maintenance, including pruning and watering, ensures the health and longevity of urban trees, maximizing their cooling potential.

6. **Community Engagement**

Public Awareness Campaigns: Raising public awareness about UHIs and the benefits of LULC strategies is crucial for successful implementation.

Community-Based Initiatives: Engaging local communities in tree planting programs, green space creation, and other initiatives fosters a sense of ownership and increases the likelihood of long-term success.

By implementing these LULC management strategies, cities can effectively mitigate the impacts of UHIs, improve the quality of life for their residents, and enhance environmental sustainability.

Key Considerations

Climate Change: The impacts of climate change, such as increased temperatures and more frequent heatwaves, exacerbate UHI effects. LULC strategies must be adapted to address these challenges.

Socioeconomic Factors: Access to green spaces and the implementation of cooling strategies can vary significantly across different socioeconomic groups. Equitable distribution of these benefits is crucial.

Long-Term Planning: LULC management strategies for UHI mitigation require long-term planning and ongoing monitoring and evaluation to ensure their effectiveness.

LULC Analysis Techniques

Remote sensing technologies provide vital data for analyzing LULC and its relationship with UHIs. Satellite imagery, aerial photography, and advanced geospatial techniques enable detailed mapping and monitoring of urban landscapes.

Satellite Data for LULC Analysis

- **Landsat Series**: The Landsat program, managed by NASA and the USGS, has provided continuous satellite imagery since 1972, making it invaluable for long-term LULC change studies. Widely utilized for analyzing built-up areas, vegetation indices, water bodies, and thermal infrared patterns. The Landsat series offers several advantages, including its moderate spatial resolution of 30 m, which makes it well-suited for regional and local studies. Its long temporal coverage, spanning decades, enables comprehensive trend analysis of Land Use and Land

Cover changes over time. Additionally, the availability of thermal infrared bands in Landsat 5, 7, 8, and 9 allows for accurate surface temperature mapping, making it a valuable tool for environmental monitoring and Urban Heat Island (UHI) studies.

- **Moderate Resolution Imaging Spectroradiometer (MODIS)**: MODIS, operated by NASA, is a vital tool for tracking dynamic environmental changes due to its high temporal resolution, capturing imagery daily. It plays a significant role in monitoring large-scale Urban Heat Island (UHI) patterns, analyzing seasonal and inter-annual LULC dynamics, and assessing vegetation health using indices such as the Normalized Difference Vegetation Index (NDVI) and Enhanced Vegetation Index (EVI). MODIS is particularly advantageous for large-area studies because of its broad swath coverage and moderate spatial resolution, ranging from 250 m to 1 km, making it ideal for regional and global environmental assessments.
- **Sentinel**: The Sentinel series of satellites, part of the European Space Agency's Copernicus Program, offers powerful tools for Land Use and Land Cover (LULC) analysis and Urban Heat Island (UHI) studies. Sentinel-1 provides radar imagery for mapping built-up areas, detecting changes in land use, and identifying impervious surfaces. Sentinel-2 offers high-resolution multispectral data, ideal for detailed land cover classification and monitoring vegetation, water bodies, and urban sprawl. Sentinel-3, equipped with a thermal sensor, directly measures Land Surface Temperature, making it crucial for analyzing UHI patterns. The advantages of the Sentinel series include frequent revisit times, high spatial resolution (down to 10 m), and an open-access data policy, ensuring widespread availability and enabling near real-time monitoring for environmental and urban studies.

Geospatial Techniques

GIS tools allow for the spatial overlay and analysis of LULC data. They enable the identification of patterns and correlations between urban growth and environmental changes.

- **Supervised and Unsupervised Classification**: Supervised and unsupervised classification are fundamental techniques in remote sensing for Land Use and Land Cover (LULC) analysis. Supervised classification relies on user-defined training data to categorize land cover types, utilizing algorithms such as Maximum Likelihood Classifier (MLC), Support Vector Machines (SVM), and Random Forest (RF). This method is highly effective for distinguishing between specific LULC types, such as urban, agricultural, and forested areas, especially when training data is well-defined and accurate. Conversely, unsupervised classification automatically groups pixels into clusters based on spectral similarity without the need for user-defined training data. Common algorithms for unsupervised classification include k-means clustering and Iterative Self-Organizing Data Analysis Techniques (ISODATA). This approach is ideal for preliminary analysis or when training data is unavailable, as it provides a first-pass categorization of the data, which can then be refined in subsequent steps (GIS Geography). Both methods are essential in LULC mapping, with supervised classification offering more precision when accurate data is available, and unsupervised classification providing a useful starting point for analysis. Studies have shown that supervised classification techniques, such as SVM and RF, often outperform unsupervised methods in terms of accuracy (Enderle and Weih 2005).

- **Thermal Remote Sensing**: Thermal Remote Sensing utilizes thermal infrared (TIR) data to measure Land Surface Temperature (LST), providing valuable insights into the Earth's surface energy balance. It plays a crucial role in identifying Urban Heat Island (UHI) hotspots by mapping temperature distributions across both urban and rural areas. This technique helps analyze the relationship between surface temperature and impervious surfaces, highlighting areas prone to higher temperatures due to urbanization. Additionally, thermal remote sensing is instrumental in studying the role of vegetation in mitigating UHI effects, as greener areas typically show cooler surface temperatures. Common data sources for thermal remote sensing include Landsat TIR bands and MODIS LST products, which offer reliable temperature measurements for large-scale environmental monitoring.
- **Spatial Metrics and Landscape Analysis**: Spatial metrics and landscape analysis are vital geospatial techniques used in Land Use and Land Cover (LULC) studies, particularly in the analysis of Urban Heat Island (UHI) effects. These techniques focus on understanding the spatial structure, configuration, and fragmentation of land cover types, which play a crucial role in shaping urban microclimates. Spatial metrics quantify aspects of urban morphology, such as patch size, shape, density, and fragmentation, offering valuable insights into how the arrangement of impervious surfaces, vegetation, and water bodies influences surface temperatures. By analyzing the spatial distribution and connectivity of different land covers, these metrics reveal patterns that contribute to elevated UHI intensities. Landscape analysis complements this by examining the broader composition and configuration of land uses, helping to identify areas where urbanization leads to greater thermal anomalies. Together, these methods help assess the relationship between urban form, land cover, and UHI, informing strategies to mitigate heat effects through sustainable urban planning, green infrastructure, and climate-sensitive design. They also contribute to understanding the broader impacts of urbanization on environmental processes like hydrology and climate, guiding more effective land management and sustainability efforts.

Machine Learning and Artificial Intelligence (ML and AI)

Machine Learning and Artificial Intelligence have recently become powerful tools for enhancing the accuracy of Land Use and Land Cover (LULC) classification, particularly in analyzing complex, multi-spectral datasets. Advanced algorithms, such as Random Forests and Convolutional Neural Networks (CNNs), enable the automated identification and classification of LULC types by learning complex patterns in satellite imagery. Random Forests is an ensemble learning method that aggregates the results of multiple decision trees to improve classification performance, particularly in cases with noisy or incomplete data. Convolutional Neural Networks (CNNs), a type of deep learning model, excel in recognizing spatial hierarchies and patterns in images, making them highly effective for processing high-dimensional, multi-spectral data and extracting nuanced features from urban, agricultural, and natural landscapes. These machine learning techniques enhance the ability to distinguish between similar land cover types, reduce classification errors, and provide more accurate and detailed LULC maps, which are critical for urban planning, environmental monitoring, and UHI studies.

Metrics for Analysis

Metrics for Analysis are essential in understanding the relationship between land cover changes and environmental properties. Key indices include the Normalized Difference Vegetation Index (NDVI), which quantifies vegetation cover, and the Normalized Difference Built-up Index (NDBI), used to identify urban areas. Land Surface Temperature (LST) measures surface heat, crucial for studying Urban Heat Island (UHI) effects. The Normalized Difference Water Index (NDWI) helps track water bodies and their cooling effects, while Urban Heat Island Intensity (UHII) measures temperature differences between urban and rural areas. These metrics provide critical insights into how urbanization, vegetation, and water bodies interact to influence temperature variations and local climate conditions.

2 Conclusions

The interplay between urbanization, LULC changes, and the UHI effect reveals critical challenges for sustainable urban development, particularly in rapidly urbanizing regions. Satellite-based remote sensing tools such as MODIS, Landsat, and Sentinel, provide robust methodologies for detecting and analyzing these changes. LULC transformations, characterized by the proliferation of impervious surfaces and the reduction of green spaces, alter energy balances, amplify surface temperatures, and exacerbate the UHI phenomenon. Geospatial techniques enhance the precision of LULC classification and UHI analysis, enabling informed decision-making. This research highlights the importance of monitoring urban expansion to mitigate adverse environmental impacts and achieve sustainable urban growth through evidence-based policy interventions and proactive planning.

References

Anitha V, Devi MP, Prabha D (2023) Bi-temporal analysis of vegetation index on land surface temperature in Kottayam, Kerala. Curr World Environ 18(3):1065–1083

Anitha V, Prabha D (2024) Role of NDVI in assessing surface urban heat island phenomenon: a case study of Alappuzha district, Kerala. J Mater Environ Sci 15(9):1331–1346

Bellard C, Bertelsmeier C, Leadley P, Thuiller W, Courchamp F (2012) Impacts of climate change on the future of biodiversity. Ecol Lett 15(4):365–377

Dutta S, Guchhait SK (2022) Assessment of land use land cover dynamics and urban growth of Kanksa Block in Paschim Barddhaman District, West Bengal. GeoJournal 87(2):971–990

Enderle DI, Weih RC Jr (2005) Integrating supervised and unsupervised classification methods to develop a more accurate land cover classification. J Ark Acad Sci 59(1):65–73

Imran HM, Hossain A, Islam AS, Rahman A, Bhuiyan MAE, Paul S, Alam A (2021) Impact of land cover changes on land surface temperature and human thermal comfort in Dhaka city of Bangladesh. Earth Syst Environ 5:667–693

Kavathekar, V., Tripathy, A. K., Chettri, S. K., & Bhanage, V. (2024) Evaluation of land use land cover dynamics and urban heat island effects over Mumbai metropolitan Region, India. Int J Environ Sci Technol 1–24

Lambin EF, Geist HJ (eds) (2008) Land-use and land-cover change: local processes and global impacts. Springer Science and Business Media

Mahato S, Pal S (2018) Changing land surface temperature of a rural Rarh tract river basin of India. Remote Sens Appl: Soc Environ 10:209–223

Melese SM (2016) Importance of non-timber forest production in sustainable forest management, and its implication on carbon storage and biodiversity conservation in Ethiopia. Int J Biodivers Conserv 8(11):269–277

Peng S, Piao S, Ciais P, Friedlingstein P, Ottle C, Bréon FM, Myneni RB (2012) Surface urban heat island across 419 global big cities. Environ Sci Technol 46(2):696–703

Pramanik S, Punia M (2020) Land use/land cover change and surface urban heat island intensity: source–sink landscape-based study in Delhi, India. Environ Dev Sustain 22:7331–7356

Sayão VM, dos Santos NV, de Sousa Mendes W, Marques KP, Safanelli JL, Poppiel RR, Demattê JA (2020) Land use/land cover changes and bare soil surface temperature monitoring in southeast Brazil. Geoderma Reg 22:e00313

Schwarz N, Lautenbach S, Seppelt R (2011) Exploring indicators for quantifying surface urban heat islands of European cities with MODIS land surface temperatures. Remote Sens Environ 115(12):3175–3186

Spångmyr M (2010) Global effects of albedo change due to urbanization. Lunds universitets Naturgeografiska institution-Seminarieuppsatser

Tesfamariam S, Govindu V, Uncha A (2023) Spatio-temporal analysis of urban heat island (UHI) and its effect on urban ecology: the case of Mekelle city, Northern Ethiopia. Heliyon 9(2)

Wemegah CS, Yamba EI, Aryee JN, Sam F, Amekudzi LK (2020) Assessment of urban heat island warming in the greater accra region. Sci Afr 8:e00426

Yang J, Gong P, Fu R, Zhang M, Chen J, Liang S, Dickinson R (2013) The role of satellite remote sensing in climate change studies. Nat Clim Chang 3(10):875–883

Zhang X, Steeneveld GJ, Zhou D, Duan C, Holtslag AA (2019) A diagnostic equation for the maximum urban heat island effect of a typical Chinese city: a case study for Xi'an. Build Environ 158:39–50

Assessing the Impact of LULC on Land Surface Temperature Variations Using GIS and Remote Sensing

Vinit Lambey, Divyarth Manas, Anurag Sharma, and Shashikant Verma

Abstract Rapid population growth and urbanization in India are transforming urban landscapes by replacing open, permeable spaces with heat-retaining concrete surfaces. This leads to the formation of urban heat islands (UHIs), where city temperatures are significantly higher than in surrounding rural areas. The present study analyzes the spatial and temporal variations in the Normalized Difference Vegetation Index (NDVI), Land Surface Temperature (LST), Normalized Difference Built-up Index (NDBI), and Land Use Land Cover (LULC) for Kharora town, Raipur District, from 2014 to 2024. The findings indicate a significant decline in vegetation cover, as reflected by a drop in mean NDVI from 0.38 in 2014 to 0.28 in 2024, primarily due to urban expansion and deforestation. Simultaneously, the LST analysis shows a rise in maximum temperature from 44.52 °C to 46.31 °C, with a reduction in cooler zones, highlighting the intensification of the UHI effect. The NDBI values further confirm rapid urbanization, with an increase in built-up areas from 5.2 sq. km to 12.6 sq. km (142.3% growth). Conversely, agricultural and forest lands have declined by 20.9 and 7.2%, respectively, while barren land has expanded by 55.6%, indicating land degradation. These findings underscore the environmental consequences of rapid urbanization and emphasize the need for sustainable urban planning, including afforestation, green infrastructure, and effective land use management, to mitigate rising temperatures and maintain ecological balance in Kharora town Raipur District, India.

V. Lambey · S. Verma
Department of Civil Engineering, National Institute of Technology, Raipur, India
e-mail: vinitlambey39@gmail.com

S. Verma
e-mail: shashiv50@gmail.com

D. Manas
Department of Civil Engineering, Shri Shankaracharya Institute of Professional Management and Technology, Raipur, India
e-mail: divyarth.manas@ssipmt.com

A. Sharma (✉)
Department of Civil Engineering, OP Jindal University, Raigarh, India
e-mail: anurag.sharma@opju.ac.in

B. Kozhikkodan Veettil and B. Das (eds.), *Remote Sensing of Urban Heat Islands*, Cities and Nature, https://doi.org/10.1007/978-3-032-11580-5_6

Keywords Urban heat islands (UHIs) · Sustainable urbanization · Heat mitigation strategies · Energy-efficient cities · Urban climate XE "Urban climate Adaptation · Green infrastructure

1 Introduction

By 2050, over 60% of the global population is projected to live in urban areas (UN 2014). This rapid urbanization is transforming land use patterns and significantly affecting environmental conditions, particularly through the intensification of the Urban Heat Island (UHI) effect. The UHI effect leads to higher temperatures in urban areas compared to nearby rural regions due to factors such as heat-retaining building materials, human-made heat emissions, and dense urban structures that restrict airflow. The effects of UHI include increased energy consumption for cooling (Mohajerani et al. 2017), greater public health risks, and reduced urban liveability (Zhang et al. 2014). As urban expansion and climate change persist, these challenges are likely to intensify, requiring further research on the severity, spatial distribution, and mitigation strategies of UHI. Land Surface Temperature (LST) is a key parameter in analyzing UHI patterns (Hachem et al. 2012), typically measured at meteorological stations (Lu et al. 2002). However, the limited distribution of these stations often leads to a generalized representation of temperature variations. This limitation is increasingly being addressed through advancements in remote sensing technologies, particularly satellite imagery (Krishna 1972; Li et al. 2013), which offer comprehensive spatial and temporal assessments of urban thermal environments.

Various spectral indices have been widely used in Urban Heat Island (UHI) studies to analyze land surface temperature (LST) variations and their relationship with land cover characteristics (Alexander 2020; Chen et al. 2006). The Normalized Difference Vegetation Index (NDVI) is one of the most used indices, derived from red and near-infrared (NIR) reflectance values, to assess vegetation density and its cooling effects in urban areas (Grover and Singh 2015; Chanpichaigosol et al. 2025). Higher NDVI values indicate dense vegetation, which helps mitigate UHI effects by reducing surface temperatures through evapotranspiration and shading (Fadhil et al. 2023). Similarly, the Normalized Difference Built-up Index (NDBI) is employed to identify built-up areas using shortwave infrared (SWIR) and NIR bands, where higher values indicate impervious surfaces contributing to increased heat retention (Kaur and Pandey 2022; Alademomi et al. 2022). The Normalized Difference Water Index (NDWI), based on green and NIR bands, is another crucial index that helps detect water bodies, which play a significant role in moderating urban temperatures (Taloor et al. 2021). Additionally, the Land Surface Temperature–Vegetation Index (LST–VI) relationship has been extensively studied to understand how vegetation cover influences temperature variations (Przeździecki et al. 2023). Other indices, such as the Enhanced Vegetation Index (EVI), which improves sensitivity to high biomass regions, and the Bare Soil Index (BSI), used for detecting barren lands, have also been applied in UHI studies (Zhao et al. 2021; Mondal 2011). These spectral indices, when analyzed together, provide a comprehensive understanding of the spatial distribution of urban heat and the role of different land cover types in mitigating or exacerbating the UHI effect.

Understanding the relationship between UHI and land use/land cover (LULC) is essential for urban planning and environmental sustainability. Remote sensing and Geographic Information Systems (GIS) have emerged as reliable tools for analyzing urban land dynamics, offering multi-spectral, multi-resolution, and multi-temporal datasets that enhance UHI assessments (Li et al. 2019). Satellite-derived indices, such as the Normalized Difference Vegetation Index (NDVI) and Normalized Difference Built-up Index (NDBI), play a crucial role in evaluating the correlation between LST and LULC through regression analyses (Weng et al. 2006; Rasul et al. 2018). Furthermore, urban classification systems like Local Climate Zones (LCZ) facilitate standardized evaluations of thermal environments (Stewart and Oke 2012). Given the growing urbanization trends and the increasing importance of sustainable city planning, this study aims to investigate the spatial and temporal variations in LULC patterns, assess UHI effects using Landsat 8 data, analyze LST concerning NDVI, NDBI, and LULC, and develop mitigation strategies to alleviate UHI impacts in Kharora town for the years 2014 and 2024.

2 Materials and Methods

2.1 Study Site

Kharora is a town in the Raipur district of Chhattisgarh, India, at around 21.4126° N and 81.9200° E (Fig. 1). It is located northeast of Raipur, the state capital, and acts as a major administrative and commercial hub for the surrounding rural districts. The town limit includes a well-defined municipal area with a mix of residential, business, and agricultural zones. Kharora's influence spreads to other adjacent communities, promoting their economic and social development. Roads, healthcare facilities, educational institutions, markets, and financial services all contribute to the region's connectedness and access to necessities. The town is well-connected by road networks to Raipur and other important areas, which promotes trade and communication. Furthermore, Kharora benefits from agricultural operations, as farming is a major vocation among the local population.

LULC changes in Kharora have had a substantial impact on local climate patterns as well as environmental sustainability. Rapid urbanization, agricultural intensification, and industrial activity have transformed the natural landscape, resulting in higher surface temperatures, less green cover, and changes in water availability (Fig. 2). The conversion of agricultural land and wood to built-up regions has exacerbated the urban heat island effect, making summers even hotter. Furthermore, deforestation and soil erosion have hampered groundwater recharge, affecting both agricultural and drinking water availability. Sustainable land use planning, afforestation, and

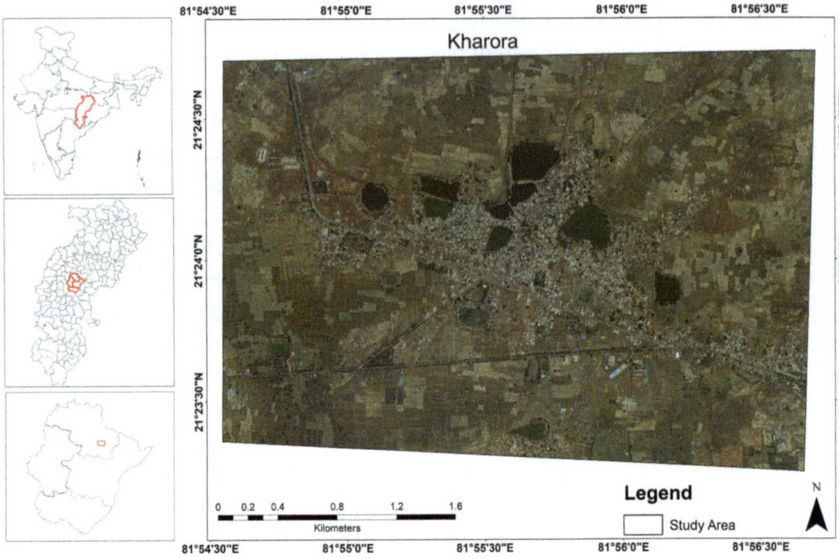

Fig. 1 Index map of the study area

sustainable urban development methods are required to counteract these effects and maintain environmental balance in the region.

We evaluated all available datasets and obtained the Landsat images from the USGS Earth Explorer platform (https://earthexplorer.usgs.gov). Table 1 outlines the specifications and characteristics of the datasets used in this study. However, the detailed information about Landsat 8 satellite imagery is shown in Table 2.

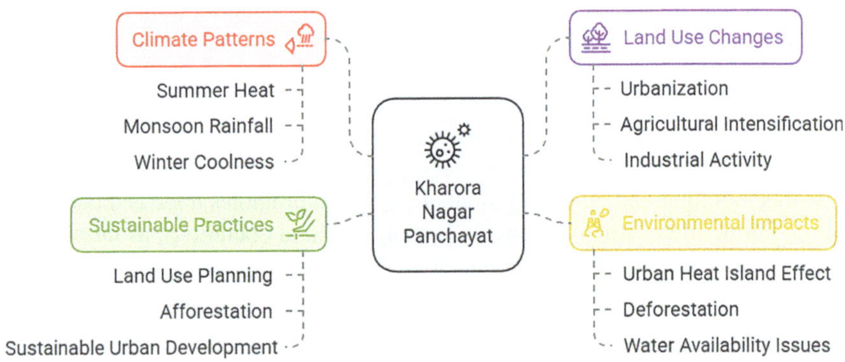

Fig. 2 Land use dynamics over the study area

Table 1 Detailed information about Landsat 8 satellite imagery

Parameters	Details
Sensor	Operational land imager (OLI) and thermal infrared sensor (TIRS)
Swath (km)	185
Spectral resolution (μm)	OLI: 0.43–2.29 (9 bands), TIRS: 10.60–12.51 (2 bands)
Spatial resolution (m)	OLI: 30 (multi-spectral), 15 (panchromatic), TIRS: 100 (resampled to 30)
Temporal resolution (days)	16
Radiometric Resolution	12-bit

Table 2 Shows the details of the datasets

Details	2014–05-05	2024–03-30
Path/row	141/45	141/45
Source	USGS	USGS
Bands comb. (NDVI)	Band 5 (NIR) and Band 4 (Red)	Band 5 (NIR) and Band 4 (Red)
Bands comb. (LST)	Band 10 (TIRS1) and Band 11 (TIRS2)	Band 10 (TIRS1) and Band 11 (TIRS2)
Bands comb. (LULC)	Band 2 (Blue), Band 3 (Green), Band 4 (Red), Band 5 (NIR), Band 6 (SWIR1), Band 7 (SWIR2)	Band 2 (Blue), Band 3 (Green), Band 4 (Red), Band 5 (NIR), Band 6 (SWIR1), Band 7 (SWIR2)

2.2 *Land Surface Temperature Estimation*

a. **Conversion of Digital Number (DN) to At-Spectral Radiance**

OLI images captured by different sensors are initially processed as 32-bit floating-point values, representing absolute radiance. In the final Level 1 product, these values are converted into 16-bit integer format. These integer values are then transformed back into spectral radiance by using the radiance scaling factors provided in the metadata file. Equation (1) can be written in ArcGIS 10.8's Band Math as follows to turn the Digital Number (DN) values of Landsat 8 Band 10 into spectral radiance:

$$L_\lambda = M_L \times Q_{CAL} + A_L \tag{1}$$

where L_λ = Spectral radiance at the sensor's aperture [W/(m^2·sr·μm)], Q_{CAL} = Quantized calibrated pixel, M_L = Multiplicative scaling factor, and A_L = Additive scaling factor.

The conversion of spectral radiance to temperature in Kelvin involves applying the inverse Planck function. This process uses the at-sensor radiance values obtained from thermal infrared sensors. The formula (Eq. 2) incorporates constants such as the sensor's calibration parameters and wavelength-specific values.

$$T_B = K_2/\ln(\frac{K_1}{L_\lambda} + 1) \quad (2)$$

where T_B = Effective brightness temperature at the sensor [K], K_1 = Calibration constant 1 [W/(m^2·sr·µm)], K_2 = Calibration constant 2 [K], L_λ = Spectral radiance at the sensor's aperture [W/(m^2·sr·µm)], and ln = Natural logarithm. However, to convert the temperature from Kelvin to Celsius, use the following formula (Eq. 3):

$$T_B = T_B - 273 \quad (3)$$

T_B = Effective at-sensor brightness temperature.

b. **Estimation of Land Surface Temperature (LST)**

LST estimation involves measuring the thermal radiation emitted from the Earth's surface using remote sensing techniques. Satellite sensors, such as those on Landsat and MODIS, capture infrared data to derive LST values (Li et al. 2023). Factors like land cover, surface emissivity, and atmospheric conditions influence these measurements. Methods such as the Radiative Transfer Equation (RTE) and split-window algorithms are commonly used for LST retrieval (Hulley et al. 2019; Zhang et al. 2022). Accurate estimation is essential for climate studies, agriculture, and urban heat analysis. By integrating ground-based validation, LST data enhances weather forecasting, drought monitoring, and environmental assessments, aiding in effective land and resource management. LST is crucial in regulating the energy exchange between land and vegetation, significantly impacting surface air temperature. The equations used to calculate LST are provided below (Eq. 4):

$$LST = \left(\frac{BT}{1}\right) + W \times (BT/14380) \times \ln(E) \quad (4)$$

where BT = Brightness temperature measured by the sensor, W = Wavelength of the emitted radiation, and E = Emissivity of the land surface.

NDVI is a vital metric for assessing vegetation health, stress, and biomass (Pettorelli 2013). It ranges from −1 to + 1, with higher values indicating healthy vegetation and negative values representing non-vegetated areas such as water, barren land, urban regions, and snow. For Landsat 7 and Landsat 8 imagery, NDVI is calculated using Eq. (5). NDBI is recognized as a key factor and indicator of LST (Malik et al. 2019). Built-up areas are particularly responsive to NDBI (Eq. 6), which ranges from −1 to + 1. Higher values indicate densely built-up regions, while lower values correspond to other land cover types.

$$\text{NDVI} = (\text{NIR} - \text{Red})/(\text{NIR} + \text{Red}) \quad (5)$$

For Landsat 8 data, NDVI = (Band 5−Band 4)/(Band 5 + Band 4)

$$\text{NDBI} = (\text{SWIR} - \text{NIR})/(\text{SWIR} + \text{NIR}) \quad (6)$$

For Landsat 8 data NDBI = (Band 6−Band 5)/(Band 6 + Band 5)

c. **Land Use Land Cover (LULC) Classification**

LULC patterns were delineated using ArcGIS software via supervised classification, utilizing the Maximum Likelihood Classification (MLC) algorithm (Mohan Rajan et al. 2020; Verma et al. 2023a). Regarded as the most precise and dependable classifier, MLC allocates pixels from an unknown class to the land use category exhibiting the highest chance of affiliation.

Four land use/land cover categories: vegetation, barren land, built-up area, and water (Verma et al. 2023b) were detected in both EMT + (2014) and OLI (2024) satellite imagery. Landsat 8 utilized bands 6, 5, and 4 for classification purposes. The area for each land use classification was determined via the Field Calculator feature in ArcGIS. Furthermore, graphical depictions of several land use categories were generated utilizing MS Excel.

3 Methodology

The methodology (Fig. 3) follows a systematic approach using USGS Landsat 7 and 8 imageries (2014 and 2024) along with topographic maps. The process starts with data downloading and study area extraction, followed by geometric rectification. Radiometric corrections include conversion to at-sensor radiance, brightness temperature, and Celsius. NDVI and LSEL estimations help derive the land surface temperature. NDVI, NDBI, and LULC analyses are used to assess urban heat island effects. A base map and Google Earth imagery validate the findings. The study ultimately evaluates spatial–temporal changes in land cover and temperature variations to understand urban heat island formation and impacts.

4 Results and Discussions

a. **Spatial and Temporal Variation of NDVI**

The NDVI analysis for Kharora town, Raipur District, shows a decline in vegetation from 2014 to 2024. The minimum, maximum, and mean NDVI values have decreased, indicating a reduction in green cover. In 2014, the mean NDVI was 0.38, reflecting a moderate decline, while by 2024, it dropped to 0.28 due to urban expansion (Table 3). The decreasing trend highlights the impact of urbanization on vegetation loss, emphasizing the need for sustainable development measures.

The evaluation of NDVI trends in Kharora town, Raipur District, indicates a steady reduction in vegetation cover over time, largely attributed to deforestation and expanding urbanization. The rise in built-up regions has led to a noticeable drop

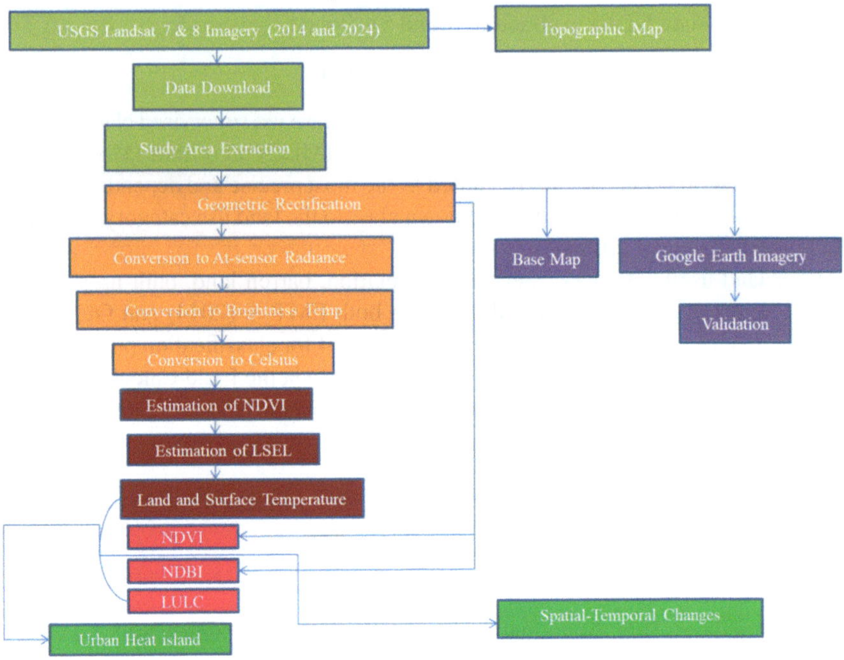

Fig. 3 Flowchart of the methodology

Table 3 Kharora town mean NDVI (temporal changes)

Year	Min NDVI	Max NDVI	Mean NDVI	Vegetation trend
2014	−0.12	0.65	0.38	Moderate decline
2024	−0.08	0.5	0.28	Urban expansion impact

in NDVI values, signifying a decline in green spaces. In contrast, rural areas exhibit NDVI fluctuations driven by agricultural activities and seasonal monsoon effects.

b. **Spatial and Temporal Variations of LST**

The LST analysis for Kharora town in 2014 (Fig. 4) reveals significant spatial temperature variations, ranging from 30.41 to 44.52 °C. The cooler regions, with temperatures between 30.41 and 35.72 °C, are primarily indicated by blue and green shades, suggesting the presence of vegetation, water bodies, or agricultural lands that help regulate temperature. In contrast, the higher temperature zones, ranging from 39.99 to 44.52 °C, are marked in red and orange, representing urbanized or barren areas that contribute to increased heat absorption, likely due to limited vegetation cover and infrastructure development. The presence of such hotspots suggests the possible impact of urban expansion and land use changes, which can intensify the urban heat island effect.

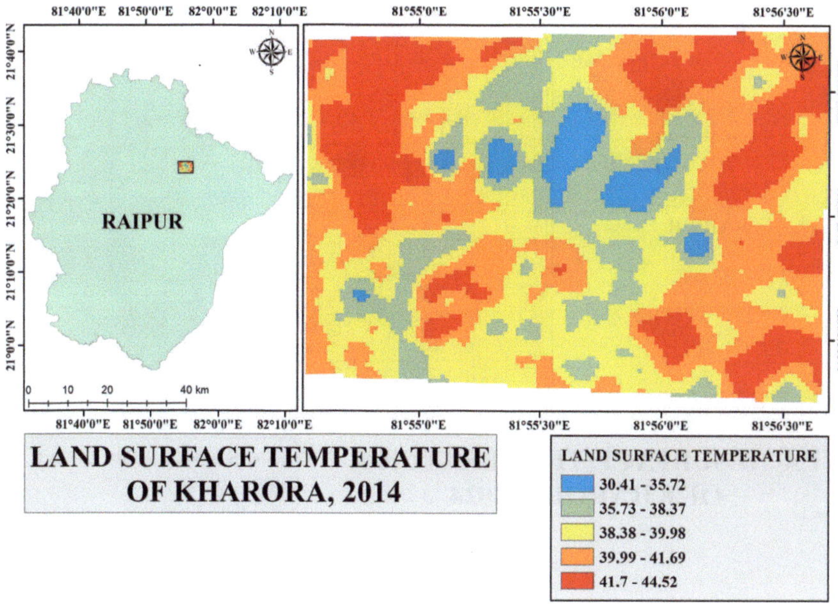

Fig. 4 Land Surface Temperature map for the year 2014

LST analysis for Kharora town in 2024 (Fig. 5) reveals a notable increase in temperature compared to 2014. The temperature now ranges between 31.06 and 46.31 °C, showing a significant rise in maximum surface temperature. The high-temperature zones, marked in red and orange (41.78–46.31 °C), have expanded, indicating rapid urbanization, infrastructure growth, and a reduction in vegetative cover. In contrast, the cooler zones (31.06–39.38 °C), shown in blue and green, have decreased, suggesting the decline of agricultural land and natural vegetation. The findings suggest an urgent need for sustainable urban planning, including afforestation, water conservation, and green infrastructure initiatives, to mitigate rising temperatures and maintain environmental stability in Kharora town.

c. **Comparison of LST from 2014 to 2024**

The LST in Kharora has undergone a significant increase from 2014 to 2024, indicating urban expansion, land use changes, and rising heat stress. The high-temperature zones (above 41 °C) have expanded, while cooler areas have reduced. Table 4 below presents a detailed comparison.

The LST analysis for Kharora town between 2014 and 2024 reveals a notable increase in overall temperature, with the maximum LST rising by 1.79 °C. The high-temperature zones (above 41 °C) have expanded significantly, indicating the intensification of the urban heat island effect. At the same time, the cooler zones (31.06–39.38 °C), which are made up of water- and plant-covered areas, have shrunk, which means there is less agricultural land and green cover. This shift is primarily driven by urbanization, deforestation, and infrastructure expansion, which contribute

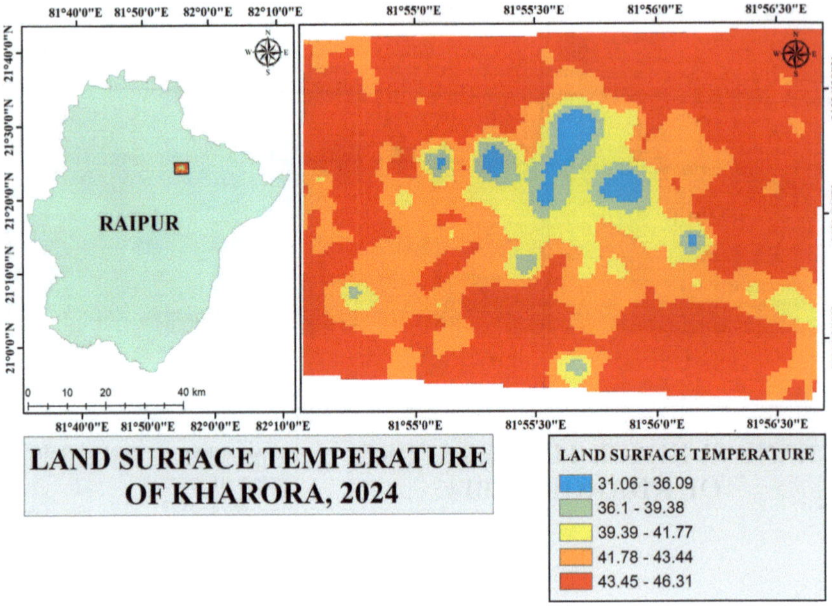

Fig. 5 Land Surface Temperature map for the year 2024

Table 4 Comparison of LST for Kharora Town (2014 vs. 2024)

Category	LST (2014)	LST (2024)	Change Observed
Minimum temperature (°C)	30.41	31.06	Slight increase (+0.65 °C)
Maximum temperature (°C)	44.52	46.31	Significant increase (+1.79 °C)
Cooler zones (Blue-Green areas, °C)	30.41–38.37	31.06–39.38	Reduced cooler zones
Hotspots (Red–Orange areas, °C)	39.99–44.52	41.78–46.31	Expansion of heat zones
High-temperature area coverage	Limited	Increased significantly	Urban heat island effect intensified
Possible causes	More vegetation and water bodies	Increased built-up areas, less green cover	Urbanization, deforestation

to greater heat absorption and retention. The increasing LST values highlight the need for sustainable land use planning.

d. **Spatial and Temporal Variations of NDBI**

The NDBI serves as a crucial metric for evaluating urbanization and the expansion of built-up areas throughout time. NDBI findings in Kharora town, Raipur district,

Chhattisgarh, help us understand how built-up surfaces change over time by looking at them in both space and time. NDBI values are derived from satellite imagery throughout various years using remote sensing techniques, illustrating the pattern of urban growth. This research is essential for assessing land use patterns, urban sprawl, and the influence of expanded built environments on temperature fluctuations and urban heat island phenomena.

The NDBI values for Kharora town demonstrate an upward trajectory over the years, signifying ongoing urban development. In 2014, the developed area was moderate, exhibiting a lower mean NDBI value of 0.15. By 2018, augmented construction and infrastructure development resulted in an elevated NDBI mean of 0.22 (Table 5). The most significant rise occurs in 2024, with the mean NDBI reaching 0.30, indicating fast urbanization and the conversion of natural land cover into developed areas. Therefore, spatial–temporal variance indicates urban sprawl, resulting in potential environmental impacts such as elevated land surface temperature, diminished green cover, and modified local climate conditions.

e. **Spatial and Temporal Variations of LULC**

The LULC patterns in Kharora town have undergone significant transformation between 2014 and 2024. A notable increase is observed in the built-up area, which expanded from 5.2 sq. km in 2014 to 12.6 sq. km in 2024, reflecting a 142.3% growth (Table 6). This increase is primarily attributed to urban expansion driven by population growth and infrastructure development. The conversion of agricultural land into residential, commercial, and industrial spaces is a major contributor to this trend. Simultaneously, agricultural land has seen a decline, reducing from 25.4 sq. km in 2014 to 20.1 sq. km in 2024, marking a 20.9% decrease. This reduction is linked to rapid urbanization, which has led to farmland being repurposed for construction and real estate projects. The shrinking agricultural land poses challenges to food security and necessitates better land management strategies. Another key observation is the slight decline in the forest area, which has shrunk from 15.3 sq. km. in 2014 to 14.2 sq. km. in 2024, amounting to a 7.2% decrease (Fig. 6). The reduction in forest cover may be due to deforestation, encroachments, or expansion of human settlements. The loss of green cover can have adverse environmental impacts, including reduced biodiversity and increased air pollution.

In contrast, the water bodies have shown a marginal increase, growing from 1.1 sq. km in 2014 to 1.3 sq. km in 2024, reflecting an 18.2% rise. This increase could be attributed to the development of artificial ponds or reservoirs to meet the water demands of the expanding population. Sustainable water resource management is

Table 5 Kharora town mean NDBI (temporal changes)

Year	Minimum NDBI	Maximum NDBI	Mean NDBI	Observation
2014	−0.15	0.45	0.15	Moderate built-up expansion
2024	−0.05	0.6	0.3	Significant urbanization and expansion

Table 6 Land use/land cover change (2014–2024)

LULC category	2014 Area (sq. km)	2024 Area (sq. km)	Change (sq. km)
Built-up area	5.2	12.6	7.4
Agricultural land	25.4	20.1	−5.3
Forest area	15.3	14.2	−1.1
Water bodies	1.1	1.3	0.2
Barren land	0.9	1.4	0.5

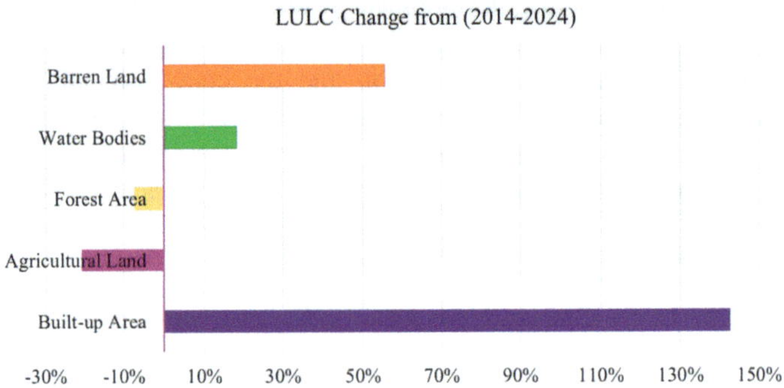

Fig. 6 LULC change from 2014 to 2024

essential to maintain this positive trend. Additionally, barren land has increased from 0.9 sq. km in 2014 to 1.4 sq. km in 2024, marking a 55.6% rise. This indicates a rise in abandoned or degraded land, which could be a result of soil erosion, deforestation, or unplanned construction activities. Addressing land degradation through afforestation and proper land use policies is crucial to prevent further expansion of barren land.

5 Conclusions and Recommendations

a. Conclusions

The study of Kharora town in Raipur District, conducted from 2014 to 2024, reveals significant environmental changes driven by urbanization. A notable decline in vegetation is evident, with the mean NDVI decreasing from 0.38 in 2014 to 0.28 in 2024, indicating a reduction in green cover due to urban expansion and deforestation. Concurrently, land surface temperature (LST) has increased, with the maximum temperature rising from 44.52 to 46.31 °C, resulting in the shrinkage of cooler zones and the expansion of high-temperature areas, thereby intensifying the urban heat island (UHI) effect. Urban expansion has been substantial, with built-up areas increasing by 142.3% (from 5.2 sq. km to 12.6 sq. km), while agricultural land has declined by 20.9% and forest cover has shrunk by 7.2%, contributing to land degradation. Additionally, barren land has expanded by 55.6%, signaling soil erosion and the loss of productive land. On a positive note, there has been a marginal increase of 18.2% in water bodies, suggesting efforts in water conservation; however, further initiatives are necessary. These findings underscore the alarming environmental impact of rapid urbanization in Kharora town. Therefore, sustainable urban planning is essential. Future strategies should prioritize afforestation, green infrastructure, sustainable land use management, and water conservation to ensure long-term environmental balance and urban sustainability in Kharora town, Raipur District.

b. **Recommendations**

The analysis findings led to several recommendations for mitigating urban heat. First, improvements in green infrastructure should be integrated into regular street upgrades and capital improvement projects to ensure ongoing progress in reducing heat across communities. Even in constrained urban spaces, small-scale green infrastructure initiatives can be effectively utilized in grassy or bare areas, vacant lots, and street rights-of-way. While urban sprawl cannot be completely halted, urban planners have sought to lessen its effects by incorporating gardens, orchards, and open fields into metropolitan corridors, allowing radiative heat to dissipate. Additionally, green roofs, or rooftop gardens, offer a valuable solution by adding a vegetative layer on building rooftops. Through evaporation, they help to lower both roof surface and ambient temperatures while functioning as natural insulators that retain heat. This decreases energy consumption for heating and cooling, improves indoor comfort, and alleviates heat stress during extreme heat events.

c. **Future scope and limitations**

This study gives us a starting point to learn more about how changes in LULC affect changes in temperature using GIS and remote sensing methods. Future research can incorporate high-resolution satellite imagery, advanced analytical models, and long-term climatic datasets to enhance precision. Expanding the study across different seasons and larger geographical regions can offer more comprehensive insights. Additionally, integrating socio-economic aspects and urban planning measures can support sustainable land management. However, certain limitations exist, such as the dependency on satellite data resolution and potential inaccuracies due to cloud cover. The study is also constrained to specific periods, limiting its temporal scope. Moreover, the absence of ground-based validation data may introduce some uncertainties. While the study focuses primarily on LULC changes, other climatic and human-induced factors affecting temperature variations warrant further investigation. Addressing these constraints can improve the accuracy and applicability of future research.

References

Alademomi AS, Okolie CJ, Daramola OE, Akinnusi SA, Adediran E, Olanrewaju HO, Alabi AO, Salami TJ, Odumosu J (2022) The interrelationship between LST, NDVI, NDBI, and land cover change in a section of Lagos metropolis, Nigeria. Appl Geomat 14(2):299–314

Alexander C (2020) Normalised difference spectral indices and urban land cover as indicators of land surface temperature (LST). Int J Appl Earth Obs Geoinf 86:102013. https://doi.org/10.1016/j.jag.2019.102013

Chanpichaigosol N, Chaichana C, Rinchumphu D (2025) Urban heat island classification through alternative normalized difference vegetation index. Glob J Environ Sci Manag 11(1):57–76. https://doi.org/10.22034/gjesm.2025.01.04

Chen XL, Zhao HM, Li PX, Yin Z (2006) Remote sensing image-based analysis of the relationship between urban heat island and land use/cover changes. Remote Sens Environ 104(2):133–146. https://doi.org/10.1016/j.rse.2005.11.016

Fadhil M, Hamoodi MN, Ziboon ART (2023) Mitigating urban heat island effects in urban environments: strategies and tools. IOP Conf Ser Earth Environ Sci 1129:012025. https://doi.org/10.1088/1755-1315/1129/1/012025

Grover A, Singh RB (2015) Analysis of urban heat island (UHI) in relation to normalized difference vegetation index (NDVI): A comparative study of Delhi and Mumbai. Environments 2(2):125–138. https://doi.org/10.3390/environments2020125

Hachem S, Duguay CR, Allard M (2012) Comparison of MODIS-derived land surface temperatures with ground surface and air temperature measurements in continuous permafrost terrain. Cryosphere 6(1):51–69. https://doi.org/10.5194/tc-6-51-2012

Hulley GC, Ghent D, Göttsche FM, Guillevic PC, Mildrexler DJ, Coll C (2019) Land surface temperature. In: Taking the temperature of the earth. Elsevier, pp 57–127. https://doi.org/10.1016/B978-0-12-814458-9.00003-4

Kaur R, Pandey P (2022) A review on spectral indices for built-up area extraction using remote sensing technology. Arab J Geosci 15(5):391. https://doi.org/10.1007/s12517-022-09688-x

Krishna R (1972) Remote sensing of urban heat islands from an environmental satellite. http://pascal-francis.inist.fr/vibad/index.php?action=getRecordDetail&idt=PASCAL731201881

Li ZL, Tang BH, Wu H, Ren H, Yan G, Wan Z, Trigo IF, Sobrino JA (2013) Satellite-derived land surface temperature: current status and perspectives. Remote Sens Environ 131:14–37. https://doi.org/10.1016/j.rse.2012.12.008

Li X, Chen WY, Sanesi G, Lafortezza R (2019) Remote sensing in urban forestry: recent applications and future directions. Remote Sensing 11(10):1144. https://doi.org/10.3390/rs11101144

Li ZL, Wu H, Duan SB, Zhao W, Ren H, Liu X, Leng P, Tang R, Ye X, Zhu J, Sun Y (2023) Satellite remote sensing of global land surface temperature: Definition, methods, products, and applications. Rev Geophys 61(1). https://doi.org/10.1029/2022RG000777

Lu D, Mausel P, Brondizio E, Moran E (2002) Assessment of atmospheric correction methods for Landsat TM data applicable to Amazon basin LBA research. Int J Remote Sens 23(13):2651–2671. https://doi.org/10.1080/01431160110109642

Malik MS, Shukla JP, Mishra S (2019) Relationship of LST, NDBI and NDVI using Landsat-8 data in Kandaihimmat watershed, Hoshangabad, India. http://nopr.niscpr.res.in/handle/123456789/45657

Mohajerani A, Bakaric J, Jeffrey-Bailey T (2017) The urban heat island effect, its causes, and mitigation, with reference to the thermal properties of asphalt concrete. J Environ Manage 197:522–538. https://doi.org/10.1016/j.jenvman.2017.03.095

Mohan Rajan SN, Loganathan A, Manoharan P (2020) Survey on land use/land cover (LU/LC) change analysis in remote sensing and GIS environment: techniques and challenges. Environ Sci Pollut Res 27(24):29900–29926. https://doi.org/10.1007/s11356-020-09091-7

Mondal P (2011) Quantifying surface gradients with a 2-band Enhanced Vegetation Index (EVI2). Ecol Ind 11(3):918–924. https://doi.org/10.1016/j.ecolind.2010.10.006

Pettorelli N (2013) The normalized difference vegetation index. Oxford University Press, USA

Przeździecki K, Zawadzki JJ, Urbaniak M, Ziemblińska K, Miatkowski Z (2023) Using temporal variability of land surface temperature and normalized vegetation index to estimate soil moisture condition on forest areas by means of remote sensing. Ecol Ind 148:110088. https://doi.org/10.1016/j.ecolind.2023.110088

Stewart ID, Oke TR (2012) Local climate zones for urban temperature studies. Bull Am Meteor Soc 93(12):879–1900. https://doi.org/10.1175/BAMS-D-11-00019.1

Taloor AK, Manhas DS, Kothyari GC (2021) Retrieval of land surface temperature, normalized difference moisture index, normalized difference water index of the Ravi basin using Landsat data. Appl Comput Geosci 9:100051. https://doi.org/10.1016/j.acags.2020.100051

UN (2014) World urbanization prospects: the 2014 revision-highlights. UN

Verma S, Kumar K, Verma MK, Prasad AD, Mehta D, Rathnayake U (2023a) Comparative analysis of CMIP5 and CMIP6 in conjunction with the hydrological processes of reservoir catchment, Chhattisgarh, India. J Hydrol Reg Stud 50:101533. https://doi.org/10.1016/j.ejrh.2023.101533

Verma S, Verma MK, Prasad AD, Mehta D, Azamathulla HM, Muttil N, Rathnayake U (2023b) Simulating the hydrological processes under multiple land use/land cover and climate change scenarios in the mahanadi reservoir complex, Chhattisgarh, India. Water 15(17):3068. https://doi.org/10.3390/w15173068

Weng Q, Lu D, Liang B (2006) Urban surface biophysical descriptors and land surface temperature variations. Photogram Eng Remote Sens 72(11):1275–1286. https://doi.org/10.14358/PERS.72.11.1275

Zhang B, Gao JX, Yang Y (2014) The cooling effect of urban green spaces as a contribution to energy-saving and emission-reduction: a case study in Beijing, China. Build Environ 76:37–43. https://doi.org/10.1016/j.buildenv.2014.03.003

Zhang X, Jiao Z, Zhao C, Qu Y, Liu Q, Zhang H, Tong Y, Wang C, Li S, Guo J, Zhu Z (2022) Review of land surface albedo: variance characteristics, climate effect and management strategy. Remote Sensing 14(6):1382. https://doi.org/10.3390/rs14061382

Zhao H, Li Y, Chen X, Wang H, Yao N, Liu F (2021) Monitoring monthly soil moisture conditions in China with temperature vegetation dryness indexes based on an enhanced vegetation index and normalized difference vegetation index. Theoret Appl Climatol 143:159–176. https://doi.org/10.1007/s00704-020-03422-x

Global Trends in Urban Heat Island Research Using Remote Sensing: A Scientometrics Analysis

Aman Kataria, Vijender Kumar Solanki, Vikram Puri, and Gloria Jeanette Rincón Aponte

Abstract Urban heat islands (UHIs) constitute an increasingly significant environmental issue, characterized by markedly higher temperatures in urban areas relative to their rural counterparts. Despite the proliferation of UHI research, a comprehensive understanding of global research trends, methodological approaches, and emerging directions in this domain remains insufficient. This study seeks to address these issues by performing a systematic scientometric analysis of UHI research utilizing remote sensing technologies. For this purpose, we have systematically collected and analyzed 300 scholarly publications focused on UHI and remote sensing. The scientometric analysis revealed three differentiated research groups, while examination of citation patterns identified key scholars whose work has been foundational to UHI research development. Keyword trends indicate that remote sensing methodologies have gained considerable prominence, particularly in recent years. Furthermore, this analysis offers a structured framework for comprehending the evolution and current state of UHI research while identifying potential future research directions.

Keywords Urban heat islands · Remote sensing · Scientometrics · Urban climate · Microclimate

A. Kataria
University Centre for Research and Development, Chandigarh University, Mohali, India
e-mail: aman.e17848@cumail.in

V. K. Solanki
Department of Computer Science and Engineering, Stanley College of Engineering and Technology For Women, Hyderabad, India

V. Puri (✉)
School of Computer Science, Duy Tan University, Danang, Vietnam
e-mail: purivikram@duytan.edu.vn

G. J. R. Aponte
Faculty of Engineering, Universidad Cooperativa de Colombia, Medellín, Colombia
e-mail: gloriaj.rincon@campusucc.edu.co

B. Kozhikkodan Veettil and B. Das (eds.), *Remote Sensing of Urban Heat Islands*, Cities and Nature, https://doi.org/10.1007/978-3-032-11580-5_7

1 Introduction

Urban heat islands (UHIs) are specific zones within urban settings that exhibit significantly elevated temperatures compared with their adjacent rural areas (Martin-Vide et al. 2015). This phenomenon generally occurs because of the high density of buildings, roads, and other infrastructure in cities, which can absorb and retain heat more efficiently than natural landscapes (Kyriakodis and Santamouris 2018). Factors such as deforestation for construction purposes and waste heat from human activities further contribute to the UHI effects. The temperature differential is predominantly observed at night as urban areas gradually release stored energy, which results in warmer nighttime conditions. These thermal disparities have diverse implications, including increased energy consumption for cooling, accelerated infrastructure deterioration, exacerbation of heat-related morbidity and mortality, and the amplification of greenhouse gas emissions (Parihar and Birman 2024).

Remote sensing has emerged as a critical tool for studying and analyzing the UHI phenomenon. With the help of satellite-enabled thermal sensors, remote sensing can capture Land Surface Temperature (LST) over a large spatial scale with high temporal resolution (Mao et al. 2021). Technologies such as Landsat, MODIS, and Sentinel satellites provide comprehensive data that enable researchers and industry experts to map UHIs (Almeida et al. 2021). This facilitates the analysis of intensity and assessment of temporal changes.

The primary advantage of these datasets is their ability to identify areas and populations vulnerable to heat by integrating socioeconomic data into heat vulnerability indices (HVI) (Niu et al. 2021). Furthermore, remote sensing offers various strategies, such as enhancing vegetation cover or redesigning urban layouts to mitigate the effects of UHI.

2 Importance of Scientometric Analysis in UHI Research

Scientometric analysis is crucial for advancing our understanding of UHI. Given that UHI research encompasses multiple disciplines, including urban planning, climate science, public health, and remote sensing, scientometrics methods offer a structured approach to analyzing the literature and its interdisciplinary connections. Figure 1 highlights the key roles of scientometrics analysis in UHI Research.

A. **Mapping Research Trends and Evolution**: Scientometric tools assist researchers in identifying emerging trends in UHI studies. For instance, citation network analysis provides insights into influential authors, institutions, and publications that have contributed significantly to the development of UHI research. Similarly, keyword co-occurrence networks have elucidated the primary focus areas of these studies.

Fig. 1 Key roles of scientometric analysis in UHI research

B. **Identifying Knowledge Gap**: Scientometric studies are instrumental in identifying underexplored areas in UHI research. Although regions such as Asia and Europe have been extensively studied using remote sensing methods, Africa remains comparatively underrepresented in this research domain (Ncongwane et al. 2021; Ngie et al. 2014).
C. **Interdisciplinary Integration**: The urban heat island (UHI) phenomenon intersects with various disciplines, including public health (e.g., heat stress), urban planning (e.g., land use), and environmental science (e.g., climate change). Scientometric analysis can uncover the interconnections among these fields and foster collaborative approaches to address the complex challenges they present.
D. **Analyzing Mitigation Strategies**: Scientometric studies assess the effectiveness of UHI by analyzing different case studies and their potential outcomes. A study demonstrated that green roofs and urban vegetation significantly reduce the temperature in local areas (Sadat et al. 2024).
E. **Policy and Decision-Making**: Scientometrics analysis offers valuable insights for urban planners and policymakers by synthesizing extensive research data pertinent to the field.

3 Applications of Remote Sensing in UHI Studies

In 1818, Luke Howard observed that urban centers, such as London, exhibited higher nocturnal temperatures than rural areas. This foundational study (Howard 1818) marked the commencement of systematic investigations into urban climate. Remote sensing technologies have significantly advanced research on UHI through the provision of extensive and high-resolution data. The primary applications of remote sensing in UHI research are as follows:

1. **Thermal Imaging**: Thermal imaging employs satellite-based sensors to measure Land Surface Temperature (LST), which is a critical metric for understanding the density and spatial distribution of UHI. Satellites such as Landsat, MODIS, and ASTER provide thermal data that facilitate the identification of urban hotspots where temperatures are significantly higher than those of the surrounding areas (Zhou et al. 2019). A mono-window algorithm was employed on Landsat data to extract LST and highlight the temperature variations between urban and rural areas (Abutaleb et al. 2015).
2. **Land Cover**: Two techniques were employed to analyze land cover: the Normalized Difference Vegetation Index (NDVI) and the Normalized Difference Built-up Index (NDBI). This comprehensive approach facilitates the examination of the relationship between land-use categories and UHI effects (Duan et al. 2025; Chanpichaigosol et al. 2025). NDVI is instrumental in quantifying vegetation density, which is directly linked to cooling effects through evapotranspiration, while NDBI aids in identifying built-up areas that contribute to heat retention.
3. **Temporal Monitoring**: Temporal monitoring utilizes satellite observations over extended periods to investigate changes in UHI dynamics (Xiong et al. 2022). This approach elucidates the seasonal variations, long-term trends, and impact of urban development on temperature patterns (ARSET 2022). A study employing MODIS data (Xiong et al. 2022) analyzed UHI trends from 2010 to 2018, facilitating the identification of periods of increasing or decreasing intensity associated with changes in land cover.
4. **Modeling and Prediction**: Remote sensing data facilitate the application of advanced modeling techniques to predict future UHI scenarios and assess mitigation strategies. Artificial intelligence algorithms, such as Convolutional Neural Networks (CNN), are employed to forecast urban hotspots (UHS) based on various variables, including elevation, NDVI, and NDBI (Li and Stouffs 2024). Other variables, such as population density and land cover changes, were also considered.

4 Methodology

The data collection methodology for this study on urban heat islands and remote sensing entails the systematic acquisition of 300 scholarly publications via Google Scholar and CrossRef. A Python script, utilizing the Beautiful Soup and request libraries, facilitates the extraction of comprehensive metadata from each publication, including titles, authors, publication years, venues, citation counts, and abstracts. The search parameters were precisely defined with the keywords "Urban Heat Islands Remote Sensing," focusing on publications to ensure contemporary relevance in this rapidly evolving field. After applying filtering, articles used in this study start from the year 2000.

Following data collection, the dataset was subjected to comprehensive processing and cleaning procedures to ensure quality and consistency. This process involves standardizing numerical data formats, eliminating duplicate entries, addressing missing values, and normalizing text fields to resolve character-encoding discrepancies. The refined dataset is subsequently stored in both CSV format for statistical analysis and BibTeX format for reference management.

4.1 Authors Analysis

Figure 2 provides key insights into author influence and citation patterns in UHI research. The left side ("Top Authors by Citation Count") reveals a stratification of scholarly impact, where researchers like JA Voogt and TR Oke demonstrate exceptional citation accumulation, positioning them as intellectual leaders whose work has fundamentally shaped the field. Their paper (Voogt and Oke 2003) on thermal remote sensing of urban climates (2003) has garnered 3,748 citations, indicating a sustained influence despite its publication age.

The right side of Fig. 2 ("Top Authors by Paper Count") complements this analysis by highlighting productivity patterns, where authors such as JA Sobrino and H Tran frequently appear. This dual visualization effectively distinguishes between citation impact and publication volume, revealing that the most-cited researchers are not necessarily the most prolific. The comparative analysis of both charts suggests the presence of distinct research strategies within the field, while others focus on fewer, highly influential publications and maintain broader research programs with more numerous contributions.

4.2 Temporal Analysis

Figure 3 reveals distinct evolutionary patterns in the UHI research keywords that reflect both the field's growth trajectory and methodological transformation. Core terminology such as "heat," "urban," and "urban heat island" which demonstrates remarkable synchronicity throughout the period, with consistent co-occurrence patterns indicating their fundamental conceptual unity within this research domain. These primary keywords exhibit pronounced growth from minimal presence in 2002–2007 (1–7 occurrences annually) to sustained prominence during 2013–2024 (16–22 occurrences), marking a substantial field expansion that corresponds with heightened

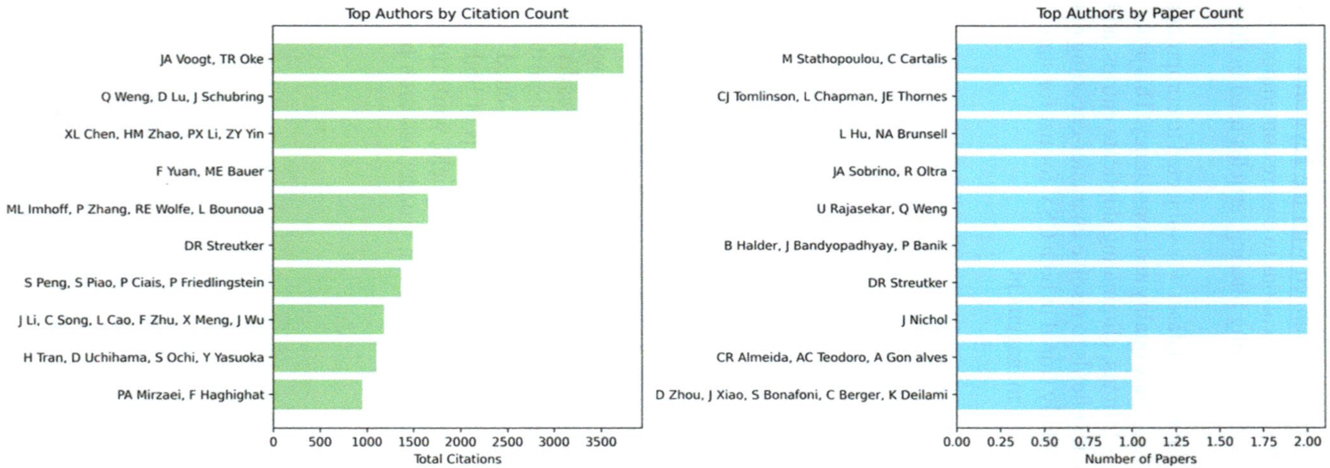

Fig. 2 Bibliometric analysis of urban heat island research showing top authors ranked by citation count (left) and publication count (right)

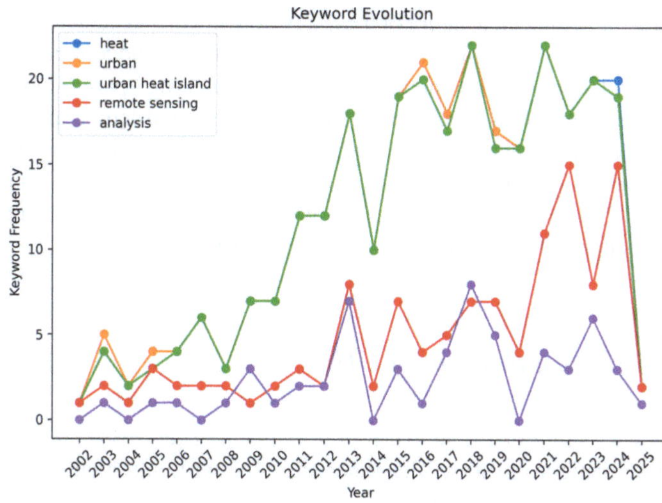

Fig. 3 Temporal evolution of keyword frequency in urban heat island research

global attention to urban climate challenges. In particular, the distinctive trajectory of "remote sensing," which maintained a relatively modest frequency (1–3 occurrences) during 2002–2012 before experiencing substantial growth beginning in 2013 and culminating in a remarkable surge during 2021–2024 that brought its frequency (15 occurrences) nearly equivalent to core terminology. This pattern signifies a fundamental methodological shift toward satellite-based observation systems, likely catalyzed by technological advancements in earth observation capabilities, improved data accessibility, and growing recognition of the efficacy of remote sensing in quantifying urban thermal patterns across diverse spatial and temporal scales.

4.3 *Keyword Analysis*

The bibliometric analysis of the UHI research literature reveals distinct keyword patterns that illuminate the field's conceptual structure and methodological foundations, as shown in Fig. 4. The frequency distribution of keywords demonstrates a clear hierarchical organization, with core terminology "heat" and "urban" (285 occurrences each) and "urban heat island" (280 occurrences) substantially outpacing all other terms, establishing them as the fundamental conceptual framework. This terminological consistency suggests a well-defined research domain with established nomenclature. A significant methodological focus becomes evident through the prominence of "remote sensing" (116 occurrences), which emerged as the dominant investigative approach, positioned far ahead of other methodological keywords. This underscores the importance of satellite-based observation techniques in contemporary UHI research. The co-occurrence patterns visualized in the heatmap further illuminate the interconnected nature of these keywords, with particularly strong associations between the core terminology and remote sensing methodologies. Secondary keywords reflect important thematic dimensions, including spatial considerations (36 occurrences), temporal dynamics (34 occurrences), and temperature measurements (39 occurrences), which characterize the multidimensional analytical frameworks employed in this field. The presence of "city" (44 occurrences) suggests an emphasis on urban scale analyses, while "satellite" (37 occurrences) reinforces the technological approaches underpinning much of this research.

4.4 Network Visualization of UHI

Figure 5 shows a bibliometric network of urban heat island research. The network displays three distinct clusters: The green cluster represents methodological and geographic research focuses, including terms like "China," "India," "impact," "effort," "evaluation," and "case study." This cluster appears to focus on research approaches and regional studies that examine the effects of UHIs.

The red cluster concentrates on measurement and analysis aspects, featuring terms like "temperature," "satellite data," "land use," "LST", "summer," and "urban heat island analysis." This group emphasizes the technical and observational components of urban heat island research.

The blue cluster contains fewer terms but includes "influence," "Shanghai," and "urban heat island effect," suggesting it may represent studies examining specific mechanisms or locations that bridge the methodological and measurement approaches.

The connections between nodes reveal important research relationships, with terms like "urban heat islands" and "urban heat island effect" having numerous links across clusters, indicating their central role in connecting different research approaches.

5 Future Directions

Remote sensing will transform UHI research in several key directions and shown in Table 1.

a. **Advanced Predictive Modeling**: Advanced artificial intelligence algorithms are currently employed to predict urban heat island (UHI) hotspots at regional scales. These models incorporate variables such as land use, vegetation indices, and population density to accurately forecast future UHI locations.

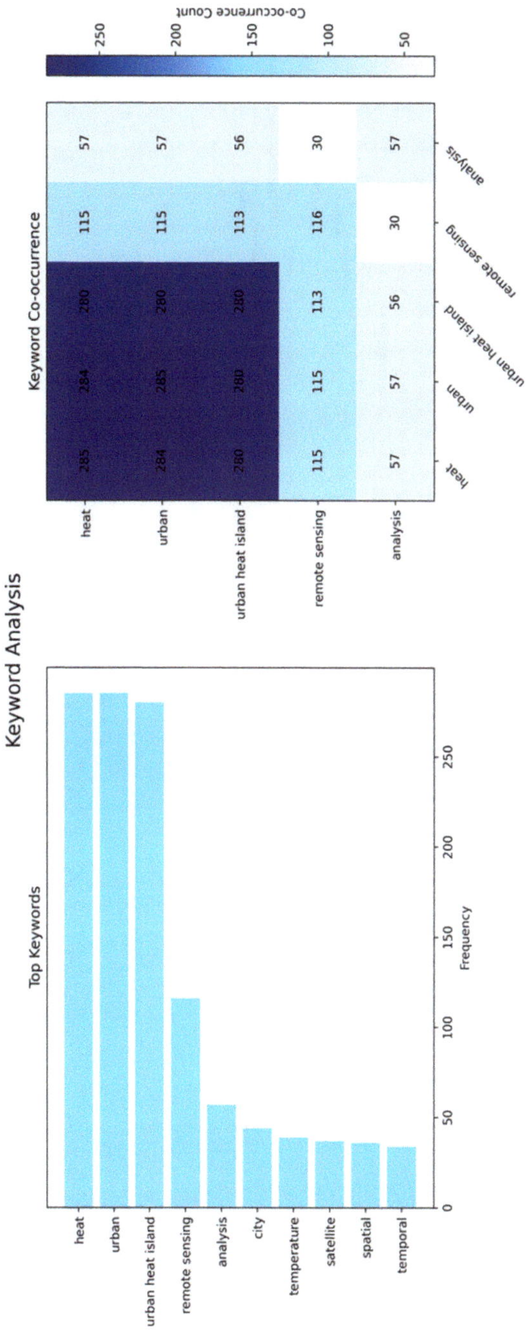

Fig. 4 Frequency distribution and co-occurrence patterns of keywords in UHI research

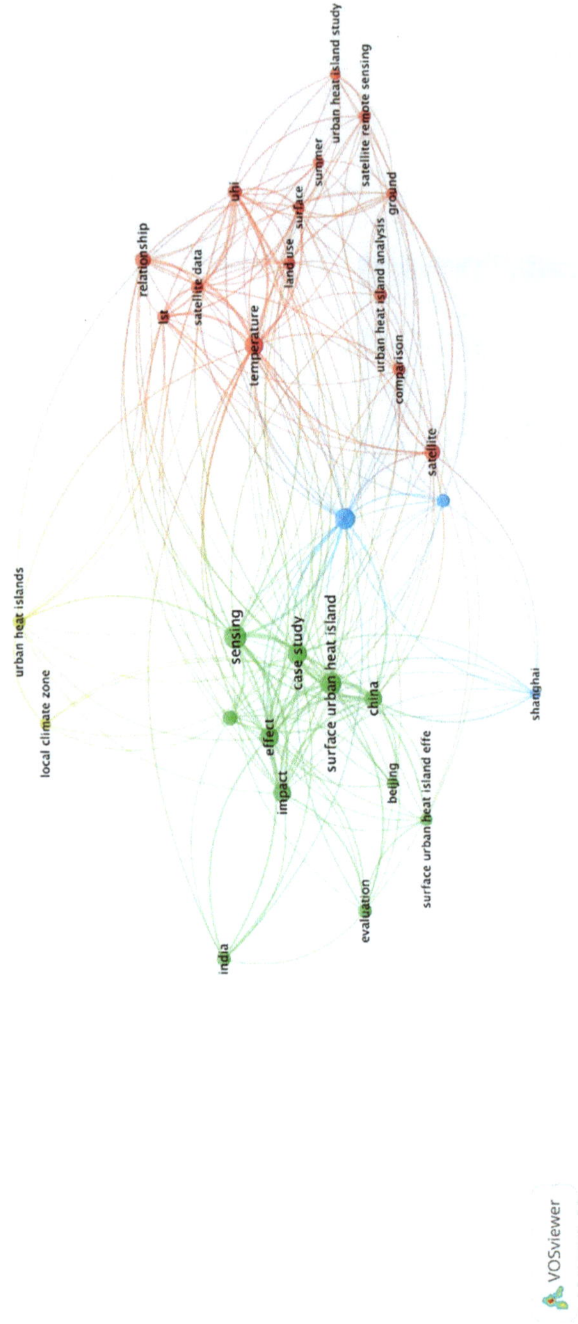

Fig. 5 Network visualization of UHI research created using VOSviewer

Table 1 Research directions in UHI

Research direction	Key variables	Expected outcomes	Applications
Advanced predictive modeling	Land use, vegetation indices, population density	Accurate forecasting of future UHI locations	Urban planning, public health interventions
Data integration	Satellite imagery, ground-based measurements, socioeconomic data, urban morphology	Comprehensive UHI assessments	More accurate UHI modeling, targeted interventions
Actionable urban planning	Evidence-based interventions, strategic green infrastructure	Clear guidelines for urban planners	UHI mitigation, improved urban livability
High-resolution temporal analysis	Daily patterns, seasonal variations	Finer temporal scale analysis of UHI dynamics	Better understanding of UHI formation and fluctuation
Climate change adaptation	Urban heat effects, climate change impacts	Understanding of interaction between UHI and climate change	Enhanced resilience planning for urban areas

b. **Data Integration**: Future research should prioritize the integration of multiple data sources, merging satellite imagery with ground-based measurements. Additionally, it should incorporate socioeconomic data and urban morphology information to facilitate comprehensive UHI assessments.
c. **Actionable Urban Planning:** Research will increasingly bridge scientific findings with practical applications, providing clear guidelines for urban planners to mitigate UHI effects through evidence-based interventions, such as strategic green infrastructure.
d. **High-Resolution Temporal Analysis**: Advances in satellite technology will enable more frequent data collection, allowing researchers to examine UHI dynamics at finer temporal scales, including daily patterns and seasonal variations.
e. **Climate Change Adaptation**: As global temperatures rise, UHI research will increasingly align with broader climate adaptation strategies, examining how urban heat effects interact with and potentially exacerbate climate change impacts in urban areas.

6 Conclusions

In this study, a scientometrics analysis of UHI research utilizing remote sensing reveals a rapidly evolving field characterized by interdisciplinary connections, methodological innovations, and increasing global significance. Our analysis highlights several key findings. First, the substantial growth in UHI publications coincides with rising global awareness of climate change impacts on urban environments and technological advancements in remote sensing capabilities. The emergence of remote sensing methodologies as fundamental to UHI research represents a transformative shift in the study of urban thermal patterns, enabling more comprehensive spatial and temporal analyses than previously possible. Second, the bibliometric network visualization reveals distinctive research clusters that underscore the multifaceted nature of UHI research, encompassing methodological approaches, measurement techniques, and regional case studies. This clustering indicates both the maturity of the field and its ongoing diversification. Third, our analysis of author influence identifies key researchers who have fundamentally shaped UHI research through highly cited contributions, while also revealing different publication strategies within the research community. This understanding of intellectual leadership provides valuable context for both established researchers and newcomers to the field. Future research on UHI utilizing remote sensing is anticipated to progress in several promising directions. These include the development of AI-powered predictive models, the integration of multi-source data, applications in evidence-based urban planning, high-resolution temporal analysis, and strategies for climate change adaptation. Such advancements will be essential in addressing the challenges posed by urban heat as cities continue to expand and global temperatures increase.

References

Abutaleb K, Ngie A, Darwish A, Ahmed M, Arafat S, Ahmed F (2015) Assessment of urban heat island using remotely sensed imagery over Greater Cairo, Egypt. Adv Remote Sens 4:35–47. https://doi.org/10.4236/ars.2015.41004

de Almeida CR, Teodoro AC, Gonçalves A (2021) Study of the urban heat Island (UHI) using remote sensing data/techniques: a systematic review. Environments 8:105. https://doi.org/10.3390/environments8100105

ARSET (2022) Satellite remote sensing for measuring urban heat islands and constructing heat vulnerability indices. NASA Applied Sciences [WWW Document]. https://appliedsciences.nasa.gov/get-involved/training/english/arset-satellite-remote-sensing-measuring-urban-heat-islands-and. Accessed on 4 May 2025

Chanpichaigosol N, Chaichana C, Rinchumphu D (2025) Urban heat island classification through alternative normalized difference vegetation index. Glob J Environ Sci Manag 11. https://doi.org/10.22034/gjesm.2025.01.04

Duan X, Haseeb M, Tahir Z, Mahmood SA, Tariq A, Jamil A, Ullah S, Abdullah-Al-Wadud M (2025) A geospatial and statistical analysis of land surface temperature in response to land use land cover changes and urban heat island dynamics. Sci Rep 15:4943. https://doi.org/10.1038/s41598-025-89167-x

Howard L (1818) The climate of London: deduced from meteorological observations, made at different places in the neighbourhood of the metropolis. W. Phillips, George Yard, Lombard Street, sold also by J. and A. Arch, Cornhill; Baldwin, Cradock, and Joy, and W. Bent, Paternoster Row; and J. Hatchard, Picadilly

Kyriakodis G-E, Santamouris M (2018) Using reflective pavements to mitigate urban heat island in warm climates - Results from a large scale urban mitigation project. Urban Clim. 24:326–339. https://doi.org/10.1016/j.uclim.2017.02.002

Li J, Stouffs R (2024) Prediction of potential urban hot spots in urban environments using convolutional neural networks and remote sensing techniques. ISPRS Ann Photogramm Remote Sens Spat Inf Sci X-4–2024, 195–202. https://doi.org/10.5194/isprs-annals-X-4-2024-195-2024

Mao Q, Peng J, Wang Y (2021) Resolution enhancement of remotely sensed land surface temperature: current status and perspectives. Remote Sens 13:1306. https://doi.org/10.3390/rs13071306

Martin-Vide J, Sarricolea P, Moreno-García MC (2015) On the definition of urban heat island intensity: the "rural" reference. Front Earth Sci 3. https://doi.org/10.3389/feart.2015.00024

Ncongwane KP, Botai JO, Sivakumar V, Botai CM (2021) A literature review of the impacts of heat stress on human health across Africa. Sustainability 13:5312. https://doi.org/10.3390/su13095312

Ngie A, Abutaleb K, Ahmed F, Darwish A, Ahmed M (2014) Assessment of urban heat island using satellite remotely sensed imagery: a review. South Afr Geogr J 96:198–214. https://doi.org/10.1080/03736245.2014.924864

Niu Y, Li Z, Gao Y, Liu X, Xu L, Vardoulakis S, Yue Y, Wang J, Liu Q (2021) A systematic review of the development and validation of the heat vulnerability index: major factors, methods, and spatial units. Curr Clim Change Rep 7:87–97. https://doi.org/10.1007/s40641-021-00173-3

Parihar J, Birman S (2024) Heat resilience in urban environments: strategies for sustainable city climate management. In: Singh P, Yadav N (eds) The climate-health-sustainability nexus: understanding the interconnected impact on populations and the environment. Springer Nature Switzerland, Cham, pp 305–324. https://doi.org/10.1007/978-3-031-56564-9_12

Sadat N, Sheikh H, Asaduzzaman M (2024) Evaluating urban heat island mitigation strategies in Rajshahi, using ENVI-Met: a remote sensing approach. J Contemp Urban Aff 8:546–556. https://doi.org/10.25034/ijcua.2024.v8n2-15

Voogt JA, Oke TR (2003) Thermal remote sensing of urban climates. Remote Sens Environ 86:370–384. https://doi.org/10.1016/S0034-4257(03)00079-8

Xiong L, Li S, Zou B, Peng F, Fang X, Xue Y (2022) Long time-series urban heat island monitoring and driving factors analysis using remote sensing and geodetector. Front Environ Sci 9. https://doi.org/10.3389/fenvs.2021.828230

Zhou D, Xiao J, Bonafoni S, Berger C, Deilami K, Zhou Y, Frolking S, Yao R, Qiao Z, Sobrino JA (2019) Satellite remote sensing of surface urban heat islands: progress, challenges, and perspectives. Remote Sens 11:48. https://doi.org/10.3390/rs11010048

Vegetation Degradation, Urban Sprawl, and Land Surface Temperature Variations in the Dalat City, Vietnam

Bijeesh Kozhikkodan Veettil and Juliana Costi

Abstract Urban sprawl in tourist destinations has resulted in the degradation of ecosystems in many areas of the world in recent decades. Unsustainable urbanization due to the pressure from population rise and booming tourism resulted in vegetation loss and an increase in impervious surfaces in Dalat City in southcentral Vietnam, which is a popular destination among national and international tourists. The present study investigated the growth of urban areas during the past five decades (1973–2022), vegetation degradation, and changes in land surface temperatures (LST) in this area. The present study used Landsat imagery series and a recent high-resolution PlanetScope data, field survey data, and statistical data. In addition, ERA5 air temperature data (2011–2022) was also used. Urban sprawl and vegetation degradation were estimated using the Normalized Difference Vegetation Index (NDVI). The present study used NDVI and Landsat thermal data for estimating LSTs of the study area. The results indicated that Dalat City underwent unprecedented and unsustainable urban growth during the study period, accompanied by vegetation loss and an increase in LST during the study period. Spatiotemporal analysis of ERA5 data indicated a continuous rise in the air temperature since 2011 with a slight decline in the same in 2020–2021. It is believed that the establishment of green corridors and environmental zoning can be effective in reducing environmental stress due to urbanization in the study area.

Keywords Mountain ecosystems · Land cover · Urbanization · Vegetation degradation · Vegetation indices

B. K. Veettil (✉)
Laboratory of Ecology and Environmental Management, Science and Technology Advanced Institute, Van Lang University, Ho Chi Minh City, Vietnam
e-mail: bkozhikkodanveettil@vlu.edu.vn

Faculty of Applied Technology, School of Technology, Van Lang University, Ho Chi Minh City, Vietnam

J. Costi
Instituto de Oceanografia, Universiade Federal Do Rio Grande, Rio Grande, Brazil
e-mail: juliana.costi@furg.br

B. Kozhikkodan Veettil and B. Das (eds.), *Remote Sensing of Urban Heat Islands*, Cities and Nature, https://doi.org/10.1007/978-3-032-11580-5_8

1 Introduction

Due to the accelerated growth of human settlements and impervious surfaces, such as roads and railways, urban spatial structure changes continuously and the uneven use of urban land is becoming severe thereby affecting the sustainable and healthy development of urban economy, society, and the environment (Cheng et al. 2021). Urban sprawl, defined as the expansion of the extend of urban areas to the outskirts of a city, creates a number of negative impacts on the environment. These negative impacts of urbanization without proper planning include deforestation and vegetation degradation, water and air pollution, and the formation of urban heat (Veettil and Grondona 2018). Among these, notable changes in spatiotemporal scales are vegetation loss/ecosystem degradation and an increase in land surface temperatures.

In many parts of the world, particularly in fast-growing urban areas in South and South Asia, a number of habitats were destroyed or fragmented due to the replacement of natural landscapes with impervious surfaces, such as concrete structures and pavements (Cao et al. 2020). Even though it is well known that widespread urbanization leads to biodiversity changes due to vegetation degradation, the relationship between urbanization and vegetation degradation is observed as complex and nonlinear, and, therefore, long-term observations are needed to understand this (Liu et al. 2022). Nevertheless, vegetation vigour and overall vegetation biomass in urban areas are considered good indicators of the impacts of urbanization on biodiversity (Hahs et al. 2009). A number of studies have used the Normalized Difference Vegetation Index (NDVI) as an effective index for the detection, mapping, and monitoring of urban vegetation and vegetation dynamics (e.g. Sun et al. 2011; Liu et al. 2015). Various vegetation parameters that can be estimated using NDVI are green leaf area, biomass, and primary (gross and net) productivity (Peng et al. 2012; Liu et al. 2015). The negative correlation between urbanization and NDVI is well established in a number of recent studies (e.g. Liu et al. 2015) and is used as an indicator of vegetation degradation in the present study.

Urbanization is usually accompanied by the replacement of natural land cover with impervious, non-evaporative surfaces (e.g. concrete buildings, asphalt roads) and these materials absorb solar radiation and store heat (Heinl et al. 2015; Mansour et al. 2022). This often results in a relatively higher land surface temperature in urbanized regions compared to the rural areas surrounding them and this phenomenon is known as urban heat islands (UHI) (Huang and Lu 2018). UHIs influence both human and environmental health in urban environments. Monitoring LST is one of the easiest methods to understand the development of UHIs from satellite data (Chen et al. 2006a; Mansour et al. 2022). A number of studies have used the correlation between LST and NDVI for understanding the complex pattern of UHI formation (e.g. Mondal et al. 2021; Guha and Govil 2022) and this method is adopted in the present study as well.

Vietnam has witnessed rapid and large-scale urbanization over the past decade which led to various social disturbances and problems related to environmental degradation (Bach and Kim 2020). Mitigating these problems is a great challenge to the

Vietnamese government in general and to the urban planning agencies in particular. Many small-scale urban areas in Vietnam in the past become highly urbanized in recent decades and the area covered by impervious surfaces has increased. The present study investigated the influence of rapid urbanization in a city in south-central Vietnam since the 1970s, and its impacts on the environment in terms of vegetation degradation and land surface temperatures. The results of the present study are believed to add light to the impacts of urbanization in mountainous terrain, particularly the variations in land surface temperatures and urban microclimate.

2 Study Site

Dalat City (11°56′30″ N–108°26′18″ E) is the capital of Lam Dong province in the central highlands of Vietnam (Fig. 1) developed approximately 12 decades ago, and has a land area of 394.38 km^2. Population in Dalat is 184,755 with a population density of 468/km^2. The annual population growth rate is about 2.5%. The town landscape of the town and architecture are in a modern style. In addition to its urbanized land cover, forests, which are dominated mainly by Pine trees, and croplands (tropical and temperate vegetables and floriculture) dominate the city's landscape.

The elevation in Dalat varies between 990 m a.s.l and 1809 m a.s.l (Nguyen et al. 2019) and the average temperature varies between 15 and 24 °C (Estoque and Murayama 2016). The dry season, which occurs between November and April, is mostly cloud-free in Dalat. Rainfall occurs between May and October in this region and a number of landslide events were reported from Dalat during this season (Thanh et al. 2020).

A number of natural and man-made lakes, including the Xuan Huong Lake in the city centre and Tuyen Lam Lake in the south of Dalat, in and around provide freshwater supply to Dalat City. Tuyen Lam Lake in its present form is due to the construction of a dam in 1987 (visible in the Landsat TM image acquired in 1988 whereas the lake is absent in the Landsat MSS image acquired in 1973). A few harmful algal bloom events in Tuyen Lam Lake were reported in local newspapers in the 2010s. Such algal blooms were believed to have related to the eutrophication of lakes by the organic materials through drainage channels coming from Dalat's urban areas.

Dalat City is a popular touristic destination located in the Langbiang Plateau (southern Central Highlands) in Vietnam, where the landscape is dominated by pine forests and tropical-temperate croplands (Veettil and Quang 2022). In fact, Dalat is a hill station developed by the French colonial government in the 1920s (Estoque and Murayama 2016). Despite being one of the most visited tourist destinations in Vietnam, more than 50% of the economic activities in Dalat City depend on agriculture, and more than 35% of the manpower in the city is involved in the agriculture-related sector. Due to this agricultural dependence, a large area within the city boundary is covered by vegetation as a part of the farms. However, there has been an unprecedented expansion of plastic-covered greenhouse farming areas

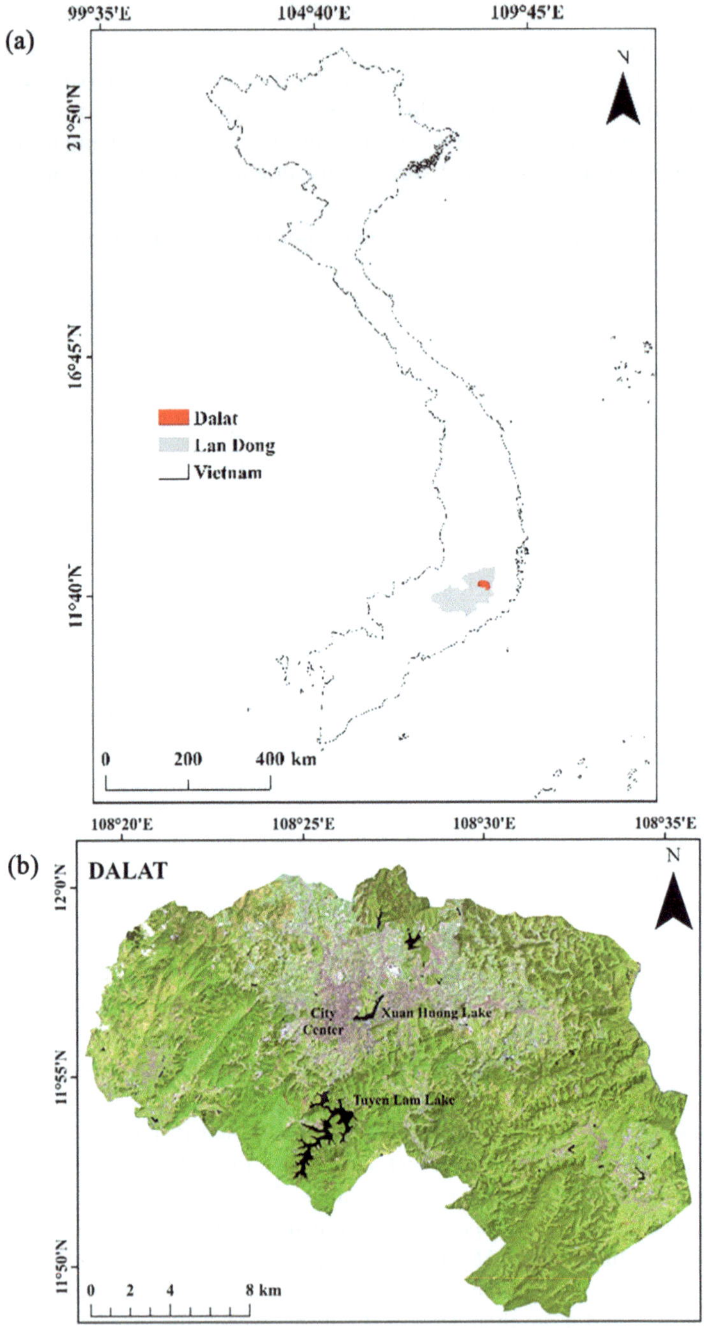

Fig. 1 **a** Geographical location of Dalat City in Vietnam, and **b** a high-resolution image (PlanetScope image subset acquired in 2022) of the study area

within and on the periphery of the city (Veettil and Quang 2022). In addition to the income from agricultural products, such as fruits and vegetables, greenhouse farms attract a large number of tourists to Dalat and have become an additional income resource in the city.

During the process of socio-economic development and urban planning, rapid urbanization and the expansion of plastic-covered greenhouse farming areas in Dalat has caused a number of problems in recent years which have led to complex environmental condition (Bach and Kim 2020; Veettil and Quang 2022). Dalat is known and recognized as 'the city of thousand flowers' and has become one of the famous brands of tourism and high-tech agriculture in the country among domestic and international visitors (Bach and Kim 2020). Moreover, the authorities have proposed the establishment of a smart city in Dalat, applying modern technologies to improve living standards in a sustainable way (Lamdong Government 2020; Bach and Kim 2020). A study by Estoque and Murayama (2016) reported that landscapes of Dalat are becoming highly fragmented due to rapid urbanization. The same study also predicted that future development faces challenges in protecting and conserving forest resources and other urban green spaces due to the high demand for built-up areas and other impervious surfaces. However, the influence of urbanization on LST using remote sensing in this region of the central highlands in Vietnam was not investigated to date. Among different environmental problems, urban flooding, deforestation, and water pollution (including algal blooms) are the major ones (Estoque and Murayama 2016).

3 Data and Methods

Landsat image series has been proven to be an effective tool in mapping land use-land cover (LULC) changes since the 1970s (Lu et al. 2012). For sustainable development and planning of urban areas, understanding the spatial patterns of land cover and its temporal changes is important (Cheng et al. 2021). In addition, satellite data has been used to estimate LST since the early 1970s (Rao 1972). The availability of medium-resolution thermal imagery, such as Landsat series (except the multispectral scanner (MSS)) and ASTER, has been utilized in a number of studies on LST measurements (Chen et al. 2006b; Zhou et al. 2019). A number of studies from Vietnam also utilized thermal satellite data since the 2010s. For example, Son et al. (2017) and Hoan et al. (2018) used Landsat data for estimating LST variations in Ho Chi Minh City and Hanoi, respectively, and Van and Bao (2020) used ASTER data in Ho Chi Minh City.

The present study used Landsat images acquired between 1973 and 2022 for understanding the spatiotemporal changes in urban areas and vegetation degradation in Dalat City. Landsat imagery can be obtained from the United States Geological Survey (USGS) at no cost (https://earthexplorer.usgs.gov/). Landsat MSS image has a spatial resolution of 60 m whereas Landsat TM and OLI data have a spatial resolution of 30 m in their optical channels (thermal bands of Landsat TM and OLI have a spatial resolution of 60 m and 100 m, respectively, resampled to 30 m). In addition to the

Landsat image series, the latest PlanetScope image acquired in 2022 was also used for understanding the detailed extent of urban settlements in Dalat City. Only Landsat TM and OLI data have thermal information and hence are used for LST changes in Dalat (Landsat MSS has only four optical bands). In addition, all the satellite imagery used in this study were acquired during the dry season (November–April) and most of the images are cloud-free during this season. Details of satellite data used for this study are given in Table 1. The expansion of urban areas is visible from the Landsat time-series data (Fig. 2) used in this study. In addition to satellite data, ERA5 air temperature data acquired between 2011 and 2022 were used to understand the recent spatiotemporal air temperature variations in the study site.

As mentioned previously, vegetation dynamics can be well studied using spectral indices, such as NDVI derived from multispectral data. NDVI is estimated as the ratio difference between measured canopy reflectance in the near-infrared and red bands (NDVI = [Near-infrared−Red]/[Near-infrared + Red]) and its value varies between −1 and + 1. A positive NDVI value indicates vegetation; the more the greenness of vegetation the more will be NDVI value and a highly negative NDVI indicates waterbodies. It is worth noting that spatial resolution is an important parameter that influences NDVI and this may cause some difficulty in comparing NDVI at different scales (Jiang et al. 2006). Nevertheless, when applied with a suitable threshold value, the NDVI method has proven its effectiveness in monitoring vegetation dynamics at high elevation (Beck et al. 2006) which is helpful in highland areas in Vietnam such as the present study site. It is worth noting that, in addition to the expansion of urban settlements and impervious surfaces, there has been an unprecedented expansion of plastic-covered greenhouse farming areas in Dalat City in recent years (Veettil and Quang 2022). It is possible that some plastic-covered greenhouse areas were misclassified as urban settlements due to similar spectral reflectance properties.

The LST measurements from satellite data became popular in the 1980s since the availability of Landsat TM with thermal information (Chen et al. 2006b; Zhou et al. 2019). LST estimates from thermal satellite data in Vietnam are not new; a number of previous studies (e.g. Hoan et al. 2018; Van and Bao 2020) estimated

Table 1 Details of landsat (MSS, TM, OLI) data used in this study

Sensor	Date of acquisition	Path/row or image id	Scene/product ID
Landsat MSS	01 January 1973	133/52	LM01_L1TP_133052_19730101_20200909_02_T2
Landsat TM	07 January 1988	124/52	LT05_L1TP_124052_19880107_20200917_02_T1
Landsat TM	02 January 1998	124/52	LT05_L1TP_124052_19980102_20200909_02_T1
Landsat TM	07 February 2011	124/52	LT05_L1TP_124052_20110207_20200823_02_T1
Landsat OLI	20 January 2022	124/52	LC08_L1TP_124052_20220120_20220127_02_T1

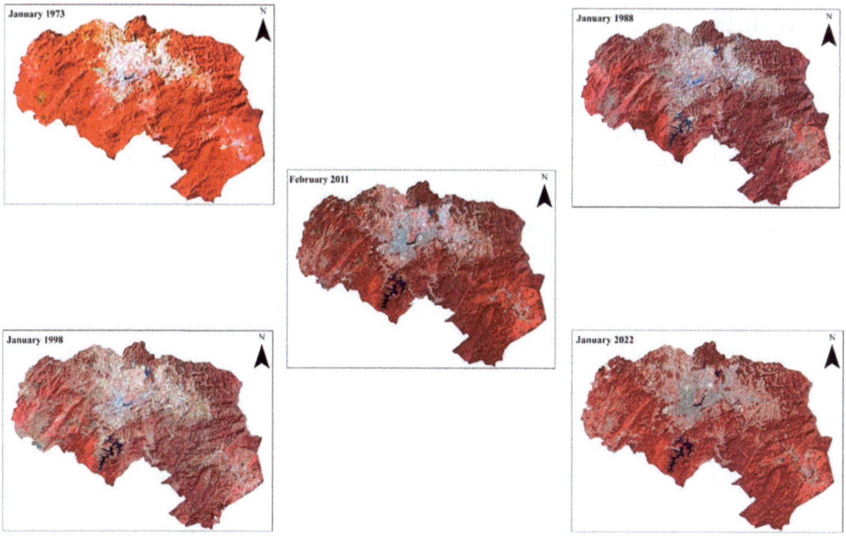

Fig. 2 Satellite images acquired in 1973, 1988, 1998, 2011, and 2022 showing the expansion of urbanized areas in Dalat

LSTs using Landsat or ASTER data. In general, LST measurements from built-up areas and impervious surfaces are higher compared to vegetation and waterbodies within urban areas (Fonseka et al. 2019). The present study used a two-step procedure to estimate LST from thermal data and NDVI from Landsat TM and OLI. In the first step, the DN values of thermal data (Band 6 of TM and Band 10 of OLI) were converted into the top of the atmosphere spectral radiance (L) as shown below:

$$L = M_L * Q_{\text{cal}} + A_L$$

where M_L represents the band-specific multiplicative rescaling factor, Qcal is the Band 10 image, and A_L is the band-specific additive rescaling factor. In the second step, L is used to estimate the brightness temperature (BT) using the formula given below:

$$\text{BT} = \{K_2/\ln\,[(K_1/L) + 1]\} - 273.15$$

where $K1$ and $K2$ are the band-specific thermal conversion constants available from the metadata. The subtraction by 273.15 is used to obtain the LST values in degrees Celsius (°C). The LST estimated in this way is significant in urban microclimate studies, particularly in the study of urban heat islands. Remote sensing is a valuable tool to estimate LST in a comparatively easy way for large areas. The estimated LST values can be further refined by using in situ measurements (Avdan and Jovanovska 2016).

4 Results

Time-series mapping of land cover in Dalat City since the 1970s indicated some serious changes over the past five decades. There has been a continuous growth of urban settlements and subsequent loss of vegetated areas in Dalat, particularly in the surroundings of the city centre, in the post-war (after 1975) period (Fig. 3). Between 1973 and 1988, it is evident that a number of old settlements were removed and new ones were constructed after the end of the Vietnam War in 1975. The renovation or construction of freshwater lakes for irrigation and drinking purposes is also visible from the LULC map in 1988. In addition, a large area of exposed or bare soil (estimated in 1973) was converted to either agricultural areas or urban settlements during the post-war period. Interestingly, there has been an expansion of urban settlements in the surroundings of Tuyen Lam Lake in the south in the early 2000s. There have been reports of harmful algal blooms in this artificial lake in 2013 due to the eutrophication of the lake by nutrients transported through drainage channels in the city. Urban sprawl in Dalat City was found to be faster between 2011 and 2022. It is worth noting that there has been a rapid increase in plastic-covered greenhouse farming areas in Dalat City since the late 2000s (Veettil and Quang 2022); plastic-covered greenhouses may also show spectral properties similar to other human-made structures and mapped as built-up areas which would explain the rapid decline in vegetated areas in recent years (2011–2022).

Estimated land surface temperatures between 1988 and 2022 showed some interesting relationships between urban sprawl and LST patterns (Fig. 4). While there

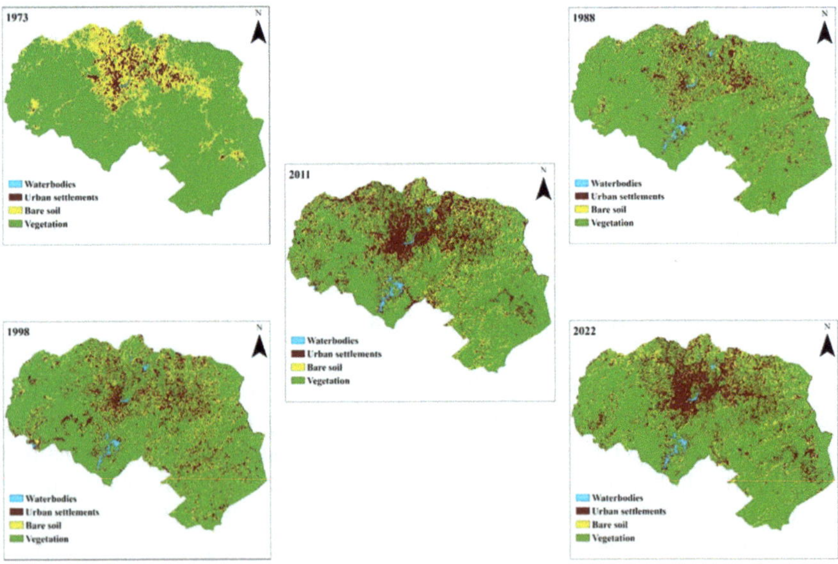

Fig. 3 Degradation of vegetated areas and urban sprawl in Dalat City (1973–2022)

was a rise in maximum LST between 1988 and 1998, it remained in the same range between 1998 and 2022 whereas there was a rise in minimum LST between 1988 and 2022. As a result of urbanization, the overall increase in minimum LST during the period was estimated at approximately 2 °C, which is significant when dealing with urban microclimate. In addition, while the highest LST values were unevenly distributed between 1988 and 1998, it is more localized in the urban areas between 2011 and 2022. Nevertheless, forest areas and waterbodies remained at lower LST throughout the study period as expected. Rapidly expanding small-scale urban areas (such as Dalat City), where the natural land cover is being replaced with impermeable and non-evaporative surfaces (e.g. concrete, asphalt) which can absorb solar energy and store heat are highly susceptible to a quick increase in LST (Heinl et al. 2015). In this perspective, understanding LST changes in rapidly growing small-scale urban areas, such as Dalat City in the Vietnamese Central Highlands, is important in understanding the development of urban heat islands in particular and urban microclimate in general.

Spatiotemporal analysis of ERA5 data (Fig. 5) showed that there has been a continuous rise in the air temperature since 2011 and there has been a slight decline in the same between 2020 and 2021. This observation in Dalat is similar to other

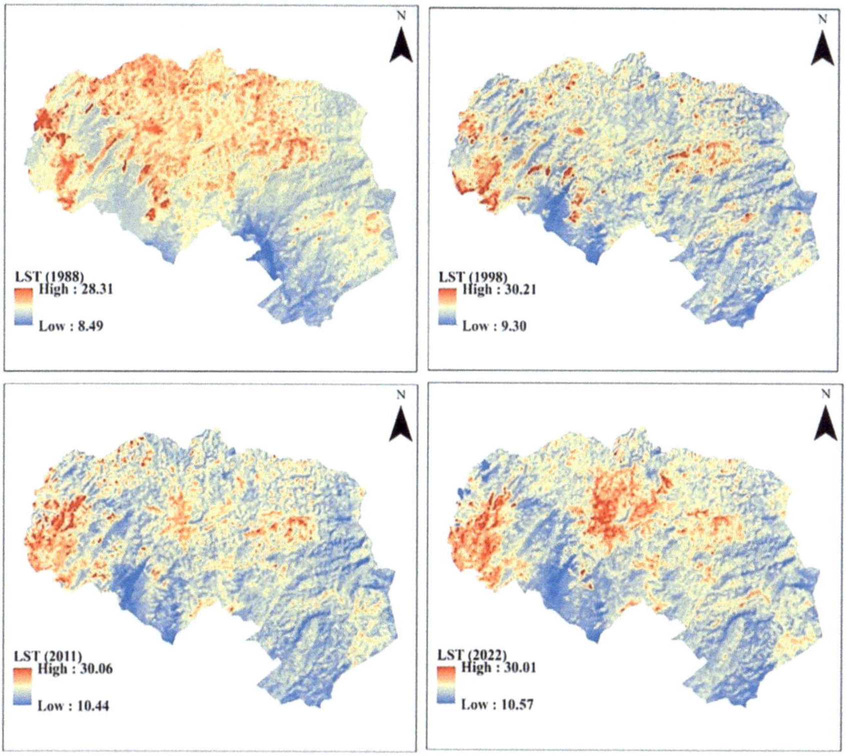

Fig. 4 Spatiotemporal changes in land surface temperatures in Dalat City (1988–2022)

urbanized areas in Vietnam (e.g. Veettil and Van 2023; Veettil et al. 2023) or elsewhere in the world (Liu et al. 2022) during the same period. It is interesting to note that this decline in the air temperature was observed during the Covid-19 pandemic lockdown period in the world, including Vietnam (Liu et al. 2022; Veettil et al. 2023).

The rural areas in Vietnam, particularly those areas having tourist destinations, have been converting to small-scale urban areas since the 2000s due to the pressure of economic development (Thi et al. 2020) and Dalat City is also not an exception to this. Such rapidly expanding small-scale urban areas are replacing the natural land cover with impervious surfaces, such as settlements, asphalt roads, and, in the case of Dalat City, a large number of plastic-covered greenhouses as well. Such areas are highly vulnerable to a rise in LST, particularly the minimum LST as observed from this study. Sustainable urbanization practices are necessary in such areas; a few studies,

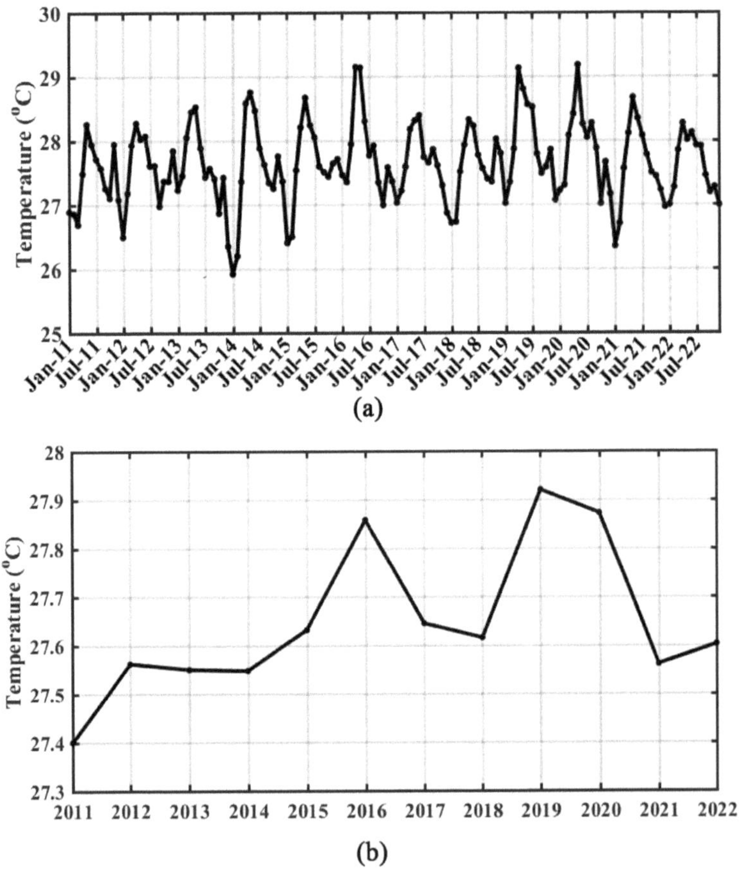

Fig. 5 **a** Average monthly air temperature (based on ERA5 data) variations between 2011 and 2023 in the three study area; **b** Average annual air temperature (based on ERA5 data) variations between 2011 and 2023 in the three study area

such as Peng et al. (2012), have underlined the importance of urban vegetation in lowering land surface temperatures in cities. However, population growth and increased tourism have resulted in the deterioration of vegetated areas in Dalat City. The establishment of the proposed smart city by the local government, despite not being an easy task, in Dalat is expected to provide some environmental benefits as well. Environmental zoning (Phan et al. 2022) can also be helpful for the sustainable development of socio-economic conditions in Dalat City.

5 Conclusions and Future Perspectives

The Central Highlands of Vietnam underwent a high rate of urbanization during the post-war period. The present study investigated the spatiotemporal changes in urban settlements in Dalat City in the central highlands of Vietnam during the past 50 years (1973–2022), the degradation of vegetation occurred during the process of urbanization, the evolution of land surface temperatures and other related environmental issues in this tourist city.

There has been a continuous growth of urban settlements and subsequent loss of vegetated areas in Dalat, particularly in the surroundings of the city centre since 1975. The present study observed that the area covered by impervious surfaces, such as urban settlements, roads, and built-up areas, has increased to a large extent between 1973 and 2022. The highest expansion of urban areas was observed in the period between 2011 and 2022. In parallel with the expansion of urbanized region, a large area of bare soil was converted to either agricultural areas or urban settlements during the post-war period. The expansion of urban settlements and increase in drainage channels transporting nutrients to the artificial lakes in the city is believed to influencing algal bloom events reported recently from Dalat.

There has been a continuous rise in the air temperature during the study period, except during the strict lockdown period during the Covid-19 pandemic between 2020 and 2021. In parallel with the rapid urbanization that occurred in the study area, there was an increase of approximately 2 °C in the minimum LST estimated in the study area whereas the maximum LST remained the same except between 1988 and 1998 when the urbanization process was at the initial stages. Air temperature data analysis also supported this finding, particularly between 2011 and 2020. Tourism is one of the major economic resources in Dalat and, therefore, improving infrastructures is also important for the economic growth of the city. The establishment of green corridors and environmental zoning can be effective in reducing environmental stress due to urbanization in the region.

References

Avdan U, Jovanovska G (2016) Algorithm for automated mapping of land surface temperature using Landsat 8 satellite data. J Sens 2016:1480307. https://doi.org/10.1155/2016/1480307
Bach KHV, Kim SK (2020) Towards evaluation the cornerstone of smart city development: case study in Dalat City, Vietnam. Smart Cities 3(1):1–16. https://doi.org/10.3390/smartcities3010001
Beck PSA, Atzberger C, Hogda KA, Johansen B, Skidmore AK (2006) Improved monitoring of vegetation dynamics at very high latitudes: a new method using MODIS NDVI. Remote Sens Environ 100:321–334. https://doi.org/10.1016/j.rse.2005.10.021
Cao Y, Wang Y, Li G, Fang X (2020) Vegetation response to urban landscape spatial pattern change in the Yangtze river delta, China. Sustainability 12(1):68. https://doi.org/10.3390/su12010068
Chen XL, Zhao HM, Li PX, Yin ZY (2006a) Remote sensing image–based analysis of the relationship between urban heat island and land use/cover changes. Remote Sens Environ 104:133–146. https://doi.org/10.1016/j.rse.2005.11.016
Chen XL, Zhao HM, Li PX, Yin ZY (2006b) Remote sensing image-based analysis of the relationship between urban heat island and land use/cover changes. Remote Sens Environ 104:133–146. https://doi.org/10.1016/j.rse.2005.11.016
Cheng L, Feng R, Wang L (2021) Fractal characteristic analysis of urban land-cover spatial patterns with spatiotemporal remote sensing images in Shenzhen City (1988–2015). Remote Sensing 13(22):4640. https://doi.org/10.3390/rs13224640
Estoque RC, Murayama Y (2016) Quantifying landscape pattern and ecosystem service value changes in four rapidly urbanizing hill stations of Southeast Asia. Landscape Ecol 31:1481–1507. https://doi.org/10.1007/s10980-016-0341-6
Fonseka HPU, Zhang H, Sun Y, Sum H, Lin H, Lin Y (2019) Urbanization and its impacts on land surface temperature in colombo metropolitan area, Sri Lanka, from 1988 to 2016. Remote Sensing 11(8):957. https://doi.org/10.3390/rs11080957
Guha S, Govil H (2022) Annual assessment on the relationship between land surface temperature and six remote sensing indices using Landsat data from 1988 to 2019. Geocarto Int 37:4292–4311. https://doi.org/10.1080/10106049.2021.1886339
Hahs AK, McDonnell M, McCarthy M, Vesk P, Corlett R, Norton B, Clemants SE, Duncan RP, Thompson K, Schwartz MW, Williams NSG (2009) A global synthesis of plant extinction rates in urban areas. Ecol Lett 12:1165–1173. https://doi.org/10.1111/j.1461-0248.2009.01372.x
Heinl M, Hammerle A, Tappeiner U, Leitinger G (2015) Determinants of urban–rural land surface temperature differences—a landscape scale perspective. Landsc Urban Plan 134:33–42. https://doi.org/10.1016/j.landurbplan.2014.10.003
Hoan NT, Liou YA, Nguyen KA, Sharma RC, Tran DP, Liou CL, Cham DD (2018) Assessing the effects of land-use types in surface urban heat islands for developing comfortable living in Hanoi City. Remote Sensing 10(12):1965. https://doi.org/10.3390/rs10121965
Huang Q, Lu Y (2018) Urban heat island research from 1991 to 2015: a bibliometric analysis. Theoret Appl Climatol 131:1055–1067. https://doi.org/10.1007/s00704-016-2025-1
Jiang Z, Huete AR, Chen J, Chen Y, Li J, Yan G, Zhang X (2006) Analysis of NDVI and scaled difference vegetation index retrievals of vegetation fraction. Remote Sens Environ 101:366–378. https://doi.org/10.1016/j.rse.2006.01.003
Lamdong Government (2020) Project of building Dalat to become a smart city in the period of 2017–2020. Office of Lamdong Provincial People's Committee, Dalat, Vietnam
Liu Y, Wang Y, Peng J, Du Y, Liu X, Li S, Zhang D (2015) Correlations between urbanization and vegetation degradation across the world's metropolises using DMSP/OLS nighttime light data. Remote Sensing 7(2):2067–2088. https://doi.org/10.3390/rs70202067
Liu Z, Lai J, Zhan W, Bechtel B, Voogt J, Quan J, Hu L, Fu P, Huang F, Li L, Guo Z, Li J (2022) Urban heat islands significantly reduced by COVID-19 lockdown. Geophys Res Lett 49:e2021GL096842. https://doi.org/10.1029/2021GL096842

Lu D, Hetrick S, Moran E, Li G (2012) Application of time series landsat images to examining land-use/land-cover dynamic change. Photogram Eng Remote Sensing 78:747–755. https://doi.org/10.14358/pers.78.7.747

Mansour K, Aziz MA, Hashim S, Effat H (2022) Impact of anthropogenic activities on urban heat islands in major cities of El-Minya Governorate, Egypt. The Egypt J Remote Sensing Space Sci 25:609–620. https://doi.org/10.1016/j.ejrs.2022.03.014

Mondal A, Guha S, Kundu S (2021) Dynamic status of land surface temperature and spectral indices in Imphal city, India from 1991 to 2021. Geomat Nat Haz Risk 12:3265–3286. https://doi.org/10.1080/19475705.2021.2008023

Nguyen VT, Tran TH, Ha NA, Ngo VL, Nadhir AA, Tran VP, Duy Nguyen HMAM, Amini A, Prakash I, Ho LS, Pham BT (2019) GIS based novel hybrid computational intelligence models for mapping landslide susceptibility: a case study at Da Lat City, Vietnam. Sustainability 11(24):7118. https://doi.org/10.3390/su11247118

Peng J, Liu Z, Liu Y, Wu J, Han Y (2012) Trend analysis of vegetation dynamics in Qinghai-Tibet Plateau using Hurst Exponent. Ecol Ind 14:28–39. https://doi.org/10.1016/j.ecolind.2011.08.011

Phan HA, Van Le T, Tran TA, Nguyen SH (2022) An environmental zoning for sustainable development in Thua Thien Hue Province, Vietnam. In: Nguyen AT, & Hens L (eds) Global changes and sustainable development in Asian emerging market economies, vol 2. Springer, Cham. https://doi.org/10.1007/978-3-030-81443-4_52

Rao PK (1972) Remote sensing of urban "heat islands" from an environmental satellite. Bull Am Meteor Soc 53:647–648

Son NT, Chen CF, Chen CR, Thanh BX, Vuong Tran H (2017) Assessment of urbanization and urban heat islands in Ho Chi Minh City, Vietnam using Landsat data. Sustain Cities Soc 30:150–161. https://doi.org/10.1016/j.scs.2017.01.009

Sun J, Wang X, Chen A, Ma Y, Cui M, Piao S (2011) NDVI indicated characteristics of vegetation cover change in China's metropolises over the last three decades. Environ Monit Assess 179:1–14. https://doi.org/10.1007/s10661-010-1715-x

Thanh DQ, Nguyen DH, Prakash I, Jaafari A, Nguyen VT, Phong TV, Pham BT (2020) GIS based frequency ratio method for landslide susceptibility mapping at Da Lat City, Lam Dong province, Vietnam. Vietnam J Earth Sci 42(1):55–66. https://doi.org/10.15625/0866-7187/42/1/14758

Thi NP, Kappas M, Wyss D (2020) Benefits and constraints of the agricultural land acquisition for urbanization for household gender equality in affected rural communes: a case study in Huong Thuy Town, Thua Thien Hue Province, Vietnam. Land 9(8):249. https://doi.org/10.3390/land9080249

Van TT, Bao HDX (2020) Study of the impact of urban development on surface temperature using remote sensing in Ho Chi Minh City, Northern Vietnam. Geogr Res 48:86–96. https://doi.org/10.1111/j.1745-5871.2009.00607.x

Veettil BK, Grondona AEB (2018) Vegetation changes and formation of small-scale urban heat islands in three populated districts of Kerala State, India. Acta Geophys 66:1063–1072. https://doi.org/10.1007/s11600-018-0189-z

Veettil BK, Quang NX (2022) Landsat-8 and Sentinel-2 data for mapping plastic-covered greenhouse farming areas: a study from Dalat City (Lam Dong Province), Vietnam. Environ Sci Pollut Res 29:73926–73933. https://doi.org/10.1007/s11356-022-20969-6

Veettil BK, Van DD (2023) Did the Covid-19 restrictions influence urban heat island phenomena in Southeast Asia? A study from Ho Chi Minh City, Vietnam. Environ Sci Pollut Res 30:66812–66821. https://doi.org/10.1007/s11356-023-26892-8

Veettil BK, Puri V, Van DD, Quang NX (2023) Variations in land surface temperatures in small-scale urban areas in Vietnam during Covid-19 restrictions: case studies from Da Nang, Hue and Vinh City. Environ Monit Assess 195:822. https://doi.org/10.1007/s10661-023-11332-4

Zhou D, Xiao J, Bonafoni S, Berger C, Deilami K, Zhou Y, Frolking S, Yao R, Qiao Z, Sobrino JA (2019) Satellite remote sensing of surface urban heat islands: progress, challenges, and perspectives. Remote Sensing 11(1):48. https://doi.org/10.3390/rs11010048

Role of Artificial Intelligence (AI) in Urban Heat Island Research

Vikram Puri and Vijender Kumar Solanki

Abstract Urban heat islands (UHIs) constitute a significant environmental issue in modern urban environments, characterized by substantially higher temperatures in urban centres compared to their adjacent rural regions. This chapter examines the transformative potential of artificial intelligence (AI) in analysing, predicting, and mitigating the UHI phenomenon. It explores how AI methodologies, including remote sensing enhanced by computer vision, deep learning architectures for geospatial temperature mapping, predictive modelling, and local climate zone classification, are advancing the understanding of urban thermal environments. Emphasis is placed on explainable AI approaches that increase transparency, elucidate critical urban factors influencing heat retention, enhance model interpretability, guide targeted interventions, and affirm model reliability. The chapter also addresses existing challenges in AI applications for UHI mitigation, such as data scarcity, the intricate nature of urban heat, dynamic urban landscapes, and scalability issues. Lastly, this chapter explores emerging technologies and research prospects, including advanced sensing networks, digital twin technology, and smart materials, which hold promise for more effective solutions to mitigate urban heat islands in an increasingly urbanized world facing the challenge of climate change.

Keywords Urban heat islands · Land cover changes · Land surface temperature · Urban planning · AI tools

V. Puri (✉)
School of Computer Science, Duy Tan University, Da Nang, Vietnam
e-mail: purivikram@duytan.edu.vn

V. K. Solanki
Stanley College of Engineering and Technology for Women, Hyderabad, India

B. Kozhikkodan Veettil and B. Das (eds.), *Remote Sensing of Urban Heat Islands*, Cities and Nature, https://doi.org/10.1007/978-3-032-11580-5_9

1 Introduction to UHI and AI

Urban heat islands (UHIs) constitute one of the foremost environmental challenges confronting contemporary urban centres. This phenomenon manifests when urban areas exhibit substantially elevated temperatures compared to their surrounding rural landscapes (Hsu et al. 2021). The temperature differential arises from a multitude of interrelated factors (Yow 2007): the extensive replacement of natural vegetation with heat-retentive materials such as concrete and asphalt; concentrated anthropogenic activities that generate waste heat; modified air circulation patterns attributable to dense architectural configurations; diminished evaporative cooling resulting from the reduction of green spaces; and the presence of heat-retaining air pollution. These elevated temperatures precipitate a series of cascading effects across urban ecosystems, which include increased energy consumption for cooling purposes, degraded air quality, compromised water systems, and escalated health risks—especially for vulnerable populations during extreme heat events (US EPA n.d.). Urban inhabitants encounter direct consequences ranging from heat-related illnesses and respiratory issues to decreased comfort and productivity, with socioeconomic disparities frequently influencing exposure levels and adaptive capacities (Meng et al. 2023). As climate change exacerbates and global urbanization progresses, addressing the UHI phenomenon has become increasingly imperative for fostering sustainable, resilient, and equitable urban environments.

The Urban Heat Island Phenomenon and Its Impacts

The urban heat island (UHI) phenomenon is manifested when urban areas exhibit elevated temperatures relative to their rural surroundings, a result of multiple interrelated factors (Phelan et al. 2015). Urban landscapes supplant natural vegetation with materials such as concrete, asphalt, and other impermeable surfaces that effectively absorb and retain heat (Shamsaei et al. 2024). Furthermore, urban human activities contribute significantly to heat emissions through transportation, industrial processes, and building operations (Rupard 2019). The architectural configuration of cities—characterized by tall structures, narrow passageways, and limited airflow—serves to entrap heat within urban confines. These heightened urban temperatures result in several critical implications. There is an escalation in energy consumption as residents and commercial entities increasingly depend on air conditioning during periods of intense heat. The UHI effect also exacerbates health risks associated with high temperatures, including heat exhaustion, heatstroke, and respiratory complications, particularly during extreme heat events. Empirical research indicates that these impacts are inequitably distributed across different population groups (Harlan et al. 2007) (Hsu et al. 2021). Low-income individuals typically experience greater urban heat exposure compared to those with higher incomes (Hsu et al. 2021). This pattern of uneven exposure underscores how the UHI effect interacts with pre-existing social and economic disparities within urban settings.

Evolution of UHI Monitoring Approaches with Artificial Intelligence

The incorporation of artificial intelligence (AI) and advanced technologies has fundamentally transformed the methodologies employed in monitoring and analysing UHIs (Mohamed and Zahidi 2024). Machine learning algorithms now facilitate the processing of extensive datasets derived from satellite imagery, meteorological stations, and Internet of Things (IoT) sensors, to identify UHI hotspots and develop predictive models (Liu et al. 2024). AI augments the examination of thermal satellite imagery, thereby enabling more precise mapping of temperature distributions and land cover classifications within urban landscapes (Gu and Zeng 2024). Predictive modelling, empowered by machine learning, allows for the forecasting of temperature changes predicated on historical datasets, thereby aiding urban planners in anticipating future impacts of UHIs. These AI applications confer substantial advantages for UHI research and management, enhancing the accuracy of UHI detection and measurement while augmenting the capabilities to process and analyse voluminous datasets. AI facilitates more precise forecasts of future UHI patterns and intensities and improves the integration of multiple data sources for holistic analysis. As urban populations continue to expand globally, the comprehension and mitigation of UHI effects assume greater importance. AI-driven methodologies furnish urban planners and policymakers (Calleo et al. 2025) with efficacious tools to address this environmental challenge, ultimately promoting the development of more sustainable and resilient urban environments. Figure 1 delineates the conceptual framework that elucidates the interrelationship between the characteristics of urban heat islands and artificial intelligence technologies. Their intersection exemplifies practical applications pertinent to solutions for urban climate management.

2 AI Methodologies for UHI Analysis and Prediction

This section provides a comprehensive discussion on various artificial intelligence methodologies employed for the analysis and prediction of urban heat islands (UHI). **Table 1** highlights UHI research based on technologies and methodologies.

Remote Sensing and Computer Vision Techniques for Thermal Pattern Recognition

The study of urban heat islands (UHI) has undergone substantial transformation by incorporating remote sensing and computer vision methodologies. These approaches employ high-resolution satellite imagery and sophisticated image processing algorithms (Liu et al. 2024; Mohamed and Zahidi 2024). Thermal infrared data obtained from platforms such as Landsat and MODIS are pivotal for acquiring land surface temperature measurements (Oh et al. 2020). When these measurements are processed using Convolutional Neural Networks (CNNs), they facilitate the automated extraction of thermal signatures within urban environments (Lau et al. 2023). These advanced computational techniques markedly improve the precision and efficiency of

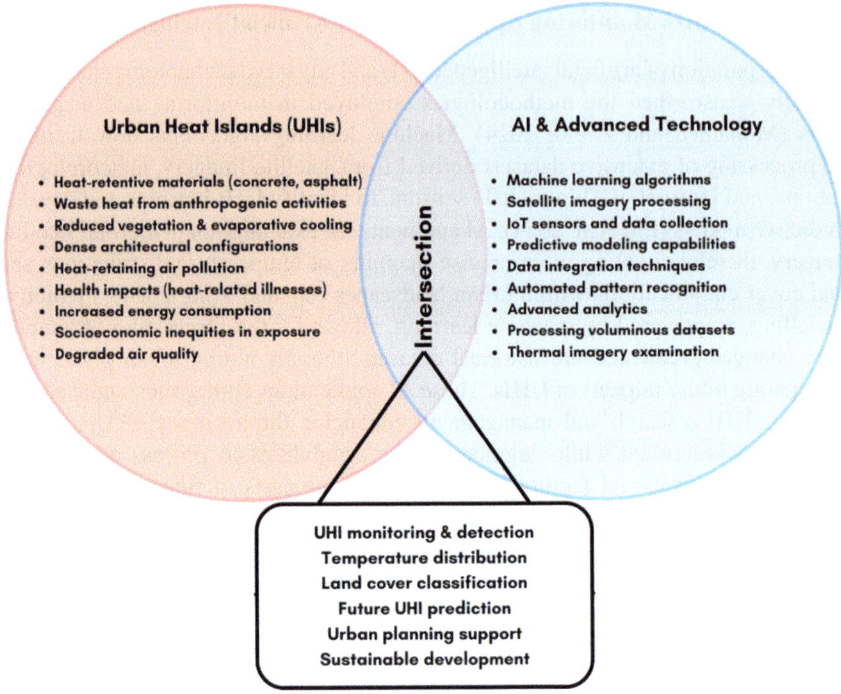

Fig. 1 Urban heat islands and AI technologies integration framework

UHI detection relative to traditional methods, thereby allowing researchers to discern subtle thermal anomalies, monitor temporal changes in urban thermal patterns, and develop more intricate understandings of the complex interactions between urban morphology and microclimate genesis.

Deep Learning Architectures for Geospatial Temperature Mapping

The utilization of deep learning architectures in geospatial temperature mapping constitutes a substantial methodological advancement within the field of urban climate studies (Tanoori et al. 2024). Long Short-Term Memory (LSTM) neural networks have exhibited exceptional proficiency in capturing the temporal evolution of urban thermal conditions, adeptly modelling the nonlinear relationships and intricate dependencies present in temperature time-series data (Zuccarini 2024). The advent of hybrid architectures that amalgamate spatial and temporal dimensions has permitted researchers to produce temperature maps with unparalleled spatial resolution while concurrently accounting for temporal dynamics (Tsokov et al. 2022; Anjos et al. 2025). These sophisticated models adeptly address the intrinsic challenges associated with urban temperature mapping: the pronounced spatial heterogeneity of urban environments and the notable temporal fluctuations in heat patterns shaped

Table 1 Urban heat island (UHI) research: technologies and methodologies

Research area	Key technologies	Applications	Techniques
Remote sensing and computer vision	– Satellite imagery – Thermal infrared data – Convolutional Neural Networks (CNNs)	– Thermal signature extraction – Detection of thermal anomalies – Monitoring temporal changes	– High-resolution image processing – Landsat and MODIS platforms
Deep learning for geospatial mapping	– LSTM neural networks – Hybrid architectures	– High-resolution temperature mapping – Temporal evolution modelling – Nonlinear relationship modelling	– Combined spatial–temporal modelling – Multi-dimensional data synthesis
Predictive modelling	– Random forest algorithms – LSTM networks – Ensemble-based methodologies	– UHI intensity forecasting – Temporal pattern recognition	– Nonlinear interaction identification – Short and long-term pattern modelling
Local climate zone classification	– Random forests – Support vector machines – Convolutional Neural Networks	–Spatial UHI distribution analysis – Land use pattern correlation – Urban thermal environment characterization	– Supervised classification –Remote sensing with morphological parameters

by diurnal cycles, seasonal variations, and meteorological conditions. By synthesizing multidimensional data streams, these methodologies facilitate more comprehensive analyses of urban heat island phenomena and furnish robust foundations for climate-responsive urban planning and mitigation strategies.

Predictive Modelling of UHI Dynamics and Temporal Evolution

The domain of predictive modelling of urban heat island (UHI) dynamics has been substantially redefined through the incorporation of advanced machine learning techniques. Random forest algorithms (Chanpichaigosol et al. 2025) have surfaced as exceptionally potent instruments for forecasting UHI intensity, proficiently identifying the intricate, nonlinear interactions between urban morphological elements (such as building density, height-to-width ratios, and surface materials) and meteorological variables. These ensemble-based methodologies markedly surpass traditional statistical models by addressing the inherent complexity and multivariate aspects of urban thermal processes. In parallel, sequence-based deep learning frameworks, particularly Long Short-Term Memory (LSTM) networks (Zuccarini 2024), have exhibited exceptional aptitude in delineating the temporal dimensionality of UHI phenomena, adeptly modelling both short-term oscillations and long-term evolutionary patterns within urban thermal settings. The memory cells embedded within these recurrent neural networks facilitate an effective representation of both proximate and remote temporal dependencies, enabling precise predictions of seasonal variations, trend developments, and prospective future scenarios under diverse urban developmental trajectories and climate change conditions.

Local Climate Zone Classification Using Machine Learning

The advent of machine learning has significantly transformed the classification process of local climate zones (LCZs), which are of paramount importance for comprehending the spatial distribution of urban heat islands (UHIs). Supervised classification methods, specifically random forests and support vector machines have been extensively employed to delineate LCZs utilizing remote sensing data in conjunction with urban morphological parameters (Bhamjee et al. 2024; Anjos et al. 2025). In recent advancements, deep learning methodologies, such as Convolutional Neural Networks (CNNs), have demonstrated substantial potential in the classification of LCZs, attaining superior accuracy and enhanced generalization across varying urban contexts (Lau et al. 2023). These sophisticated classification techniques facilitate researchers in more precisely characterizing urban thermal environments and their correlation with land use patterns.

3 Explainable AI Methods for Urban Heat Analysis

Explainable AI methodologies are transforming the comprehension of urban heat islands, facilitating the development of focused and efficacious mitigation strategies. In contrast to traditional "black box" methodologies, these transparent algorithms elucidate the precise manner in which factors such as building density, materials, and vegetation coverage affect the distribution of temperature across urban areas. Urban planners in metropolises such as Phoenix ("Street Transportation Cool Corridors Program" n.d.) and Singapore (Ruefenacht and Acero 2017) are actively utilizing these insights to design cooling corridors, optimize the placement of green spaces, and revise building codes. By quantifying the impact of various interventions, explainable AI enables municipalities to prioritize economically viable solutions—from the application of reflective roofing to the strategic planting of trees—that can lower temperatures in susceptible neighbourhoods. This data-driven approach is particularly significant for climate adaptation planning, providing cities with the ability to safeguard vulnerable populations from escalating heat waves while enhancing overall urban sustainability. The primary objectives of explainable AI (XAI) are diverse and essential for the progression of urban heat analysis and planning, as follows (see in Fig. 2).

Enhancing Transparency and Trust

XAI methodologies advance the comprehensibility of AI systems utilized in urban development and thermal mapping applications. By elucidating the opaque nature of algorithmic decision-making frameworks, explainable AI fosters trustworthiness

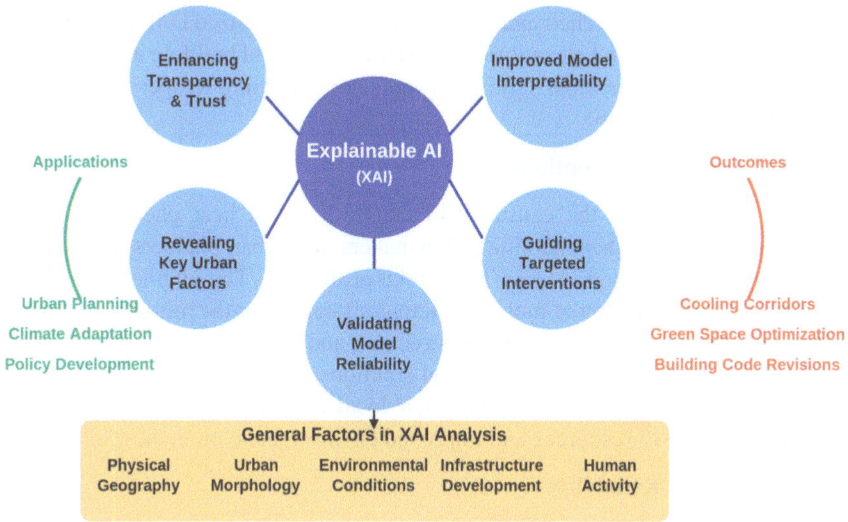

Fig. 2 Explainable AI (XAI) applications in urban heat island analysis

among various stakeholders, including municipal designers, government officials, and community members. Such transparency is pivotal in facilitating responsible governance and endorsing participatory city planning approaches that integrate diverse perspectives and cater to the heterogeneous needs of urban populations.

Revealing Key Urban Factors

Techniques of XAI, especially SHapley Additive exPlanations (SHAP), have demonstrated significant utility in elucidating the intricate interconnections between urban morphology and thermal environments. In a contemporary study (Feng et al. 2024), researchers utilized a random forest model augmented with XAI methodologies to examine the pivotal elements affecting land surface temperature (LST) patterns within China. Through the application of SHAP value analysis, the research delineated the primary factors determining LST in the urban milieu of Beijing: elevation (44.19%), compactness of impervious surfaces (17.27%), Normalized Difference Vegetation Index (11.12%), proportion of impervious surface area (8.04%), and tree height (3.83%). These outcomes provide empirical validation of the complex dynamics of urban heat in metropolitan areas, offering essential insights for urban planning and strategies for heat mitigation.

Improved Model Interpretability

In the context of urban heat island (UHI) analysis, XAI techniques are valuable for interpreting complex machine learning models. For instance, researchers (Sheng et al. 2025) studying the regional heat island effect in the Greater Bay Area employed SHapley Additive exPlanations (SHAP) values to identify key factors influencing land surface temperature. Their analysis revealed that Normalized Difference Vegetation Index (NDVI), digital elevation model (DEM) data, and building anthropogenic heat emissions were the most significant indicators affecting temperature patterns across the urban landscape.

Guiding Targeted Interventions

Through elucidation of the causative elements of urban heat phenomena, XAI provides essential insights enabling policymakers and urban planners to formulate meticulously tailored interventions (Kiouvrekis et al. 2025). This augmented comprehension aids in the creation of more efficacious strategies for the mitigation of urban heat islands, optimization of green infrastructure, and enhancement of overall urban sustainability. Instead of applying general solutions, stakeholders can focus on the specific factors most impactful within their local context, thereby ensuring more efficient allocation of resources and improved environmental outcomes.

Validating Model Reliability

XAI methodologies are instrumental in establishing the credibility of AI models within urban settings. A pertinent instance involves researchers examining the prediction of electric field strength in urban landscapes, who employed XAI approaches to elucidate the correlations between urban characteristics and electromagnetic field configurations (Kiouvrekis et al. 2025). This application elucidated how distinct elements, including population density, building height, construction materials, and spatial arrangements, exert substantial influence on the distributions of electric field strength across urban domains. By rendering these complex interdependencies comprehensible, XAI bolsters confidence in model outputs and furnishes urban planners and telecommunication engineers with practical insights for infrastructure development and signal optimization purposes.

4 Challenges and Future Directions

Challenges and Limitations of Current AI Approaches for Urban Heat Islands

Present AI methodologies for mitigating urban heat islands encounter notable obstacles and constraints. The primary challenges are as follows:

1. **Data Scarcity and Data Issues**: The performance of AI models is contingent upon the availability of high-quality, comprehensive data ("AI for Climate Action in Developing Countries: Opportunities, Challenges and Risks" n.d). Regrettably, numerous urban areas, especially within developing regions, experience substantial deficiencies in the measurement of urban heat, land use mapping, and microclimatic monitoring. This scarcity of data adversely affects the accuracy of predictive models and impairs the execution of effective urban heat island (UHI) mitigation strategies.

2. **Multifaceted Nature of UHI**: Urban heat islands arise from a complex interplay of factors such as building materials, urban geometry, vegetation coverage, and human activities, all of which contribute to this phenomenon (Yuan 2024). AI models often struggle to simultaneously account for these numerous variables and their interdependencies, risking the oversight of crucial interactions that shape urban microclimates.
3. **Dynamic Urban Landscapes**: Urban environments are perpetually undergoing transformation, manifesting through alterations in land utilization, architectural advancements, and demographic changes ("Heat in the city" n.d.). AI models are required to adapt to this urban fluidity but encounter substantial difficulties in capturing and responding to these changes in real time, which restricts their capability to deliver consistently accurate evaluations of emergent heat island effects.
4. **Scalability Issues**: Interventions successful in one urban context frequently encounter failure when applied to alternative settings, attributed to fundamental disparities in urban design, climatic conditions, and socioeconomic factors. A continual challenge in this domain is the formulation of adaptable AI methodologies (Fu et al. 2024) capable of scaling across various urban environments while preserving predictive precision and applicability.

Emerging Technologies and Research Opportunities

The development of emerging technologies and opportunities for research is progressing swiftly to mitigate urban heat islands (UHI) and their associated effects and highlighted in Table 2.

1. **Advanced Sensing and Data Collection**: State-of-the-art satellite imagery, along with burgeoning IoT sensor networks, is transforming urban heat monitoring by enabling real-time, high-resolution data capture with unparalleled levels of detail.
2. **Digital Twin Technology**: Digital twin technology facilitates an extensive analysis of UHI by amalgamating intricate three-dimensional urban models with environmental simulations. This integration empowers urban planners to observe the multifaceted interactions between urban morphology, solar radiation, aerodynamic flow patterns, and temperature distribution, thereby enabling the optimization of thermal mitigation strategies.
3. **Smart Materials and Surfaces**: UHI mitigation materials encompass solar reflective coatings that diminish heat absorption and subsequently reduce indoor temperatures in buildings with natural ventilation. Concurrently, smart glass embedded with metal oxide layers modulates its tint in response to temperature variations or solar positioning, thus facilitating effective regulation of thermal conditions within structures.

Table 2 Challenges and emerging technologies in AI-based urban heat island mitigation

Category	Key aspects	Description
Challenges and limitations	Data scarcity	Insufficient high-quality data in developing regions leading to reduced model accuracy and impaired mitigation strategy implementation
	Multifaceted UHI nature	Complex interplay of building materials, urban geometry, vegetation, and human activities that AI models struggle to capture simultaneously
	Dynamic urban landscapes	Continuous urban transformation through land use changes and development that limits real-time assessment capabilities
	Scalability issues	Context-specific interventions failing in different urban settings due to fundamental differences in design, climate, and socioeconomic factors
Emerging technologies	Advanced sensing	State-of-the-art satellite imagery and IoT sensor networks enabling real-time, high-resolution urban heat monitoring
	Digital twin technology	Integration of detailed 3D urban models with environmental simulations for optimized thermal mitigation strategies
	Smart materials	Solar reflective coatings and responsive surfaces with embedded metal oxide layers that reduce heat absorption and provide effective thermal regulation

5 Conclusions

The integration of artificial intelligence in urban heat island research and mitigation represents a notable advancement in addressing one of the foremost urban environmental challenges in this era. This chapter illustrates how AI methodologies have revolutionized the capacity to monitor, analyse, predict, and ultimately mitigate the effects of urban heat islands with exceptional precision and insight. The combination of remote sensing with computer vision techniques has facilitated automated extraction of thermal signatures and anomaly detection at high spatial resolutions. Deep learning architectures have demonstrated exceptional capability in mapping the complex spatiotemporal dynamics of urban temperatures. Predictive modelling utilizing random forests and LSTM networks has augmented the ability to forecast UHI intensity and progression, while machine learning approaches to local climate zone classification have enhanced our comprehension of the relationship between urban form and the thermal environment. Of particular significance, the advent of explainable AI methodologies has bridged the divide between intricate algorithmic outputs and practical applications in urban planning. By quantifying the contributions of specific factors such as impervious surface coverage, vegetation indices, and building configurations to urban heat formation, these transparent methodologies enable data-driven decision-making for city planners, policymakers, and community stakeholders. Despite these advancements, considerable challenges persist. Data scarcity, especially in developing regions, hampers model accuracy and applicability.

The inherent complexity of urban thermal environments continues to challenge even sophisticated AI models, while the evolving nature of urban development necessitates adaptive strategies. Moreover, the context-specific nature of successful interventions requires AI methodologies capable of scaling across varied urban environments.

References

AI for Climate Action in Developing Countries: Opportunities, Challenges and Risks [WWW Document] (n.d.) https://unfccc.int/ttclear/tec/AI4climate.html#infonote. Accessed on 14 Mar 2025

Anjos M, Medeiros D, Castelhano F, Meier F, Silva T, Correia E, Lopes A (2025) LCZ4r package R for local climate zones and urban heat islands. Sci Rep 15:7710. https://doi.org/10.1038/s41598-025-92000-0

Bhamjee M, Debary H, Gaffoor Z, Govindasamy T, Mahlasi C, Fiaz M, Vos E, Klein L, Makhanya S, Watson C, Kuehnert J (2024) Detection and characterization of urban heat islands with machine learning. In: Presented at the IGARSS 2024–2024 IEEE international geoscience and remote sensing symposium. pp 1693–1699. https://doi.org/10.1109/IGARSS53475.2024.10641750

Calleo Y, Taylor A, Pilla F, Di Zio S (2025) AI-assisted real-time spatial Delphi: integrating artificial intelligence models for advancing future scenarios analysis. Qual Quant. https://doi.org/10.1007/s11135-025-02073-2

Chanpichaigosol N, Chaichana C, Rinchumphu D (2025) Urban heat island classification through alternative normalized difference vegetation index. Glob J Environ Sci Manag 11. https://doi.org/10.22034/gjesm.2025.01.04

Feng F, Ren Y, Xu C, Jia B, Wu S, Lafortezza R (2024) Exploring the non-linear impacts of urban features on land surface temperature using explainable artificial intelligence. Urban Clim 56:102045. https://doi.org/10.1016/j.uclim.2024.102045

Fu Q, Zheng Z, Sarker MNI, Lv Y (2024) Combating urban heat: systematic review of urban resilience and adaptation strategies. Heliyon 10. https://doi.org/10.1016/j.heliyon.2024.e37001

Gu Z, Zeng M (2024) The use of artificial intelligence and satellite remote sensing in land cover change detection: review and perspectives. Sustainability 16:274. https://doi.org/10.3390/su16010274

Harlan S, Brazel AJ, Darrel Jenerette G, Jones NS, Larsen L, Prashad L, Stefanov WL (2007) In the shade of affluence: the inequitable distribution of the urban heat island. Res Soc Probl Public Policy 15:173–202. https://doi.org/10.1016/S0196-1152(07)15005-5

Heat in the city: urban heat island effect [WWW Document] (n.d.) BAU Münch. https://bau-muenchen.com/en/bau-insights/detail/heat-in-the-city-urban-heat-island-effect.html. Accessed on 14 Mar 2025

Hsu A, Sheriff G, Chakraborty T, Manya D (2021) Disproportionate exposure to urban heat island intensity across major US cities. Nat Commun 12:2721. https://doi.org/10.1038/s41467-021-22799-5x

Kiouvrekis Y, Psomadakis I, Vavouranakis K, Zikas S, Katis I, Tsilikas I, Panagiotakopoulos T, Filippopoulos I (2025) Explainable machine learning-based electric field strength mapping for urban environmental monitoring: a case study in paris integrating geographical features and explainable AI. Electronics 14:254. https://doi.org/10.3390/electronics14020254

Lau T-K, Chen Y-C, Lin T-P (2023) Application of local climate zones combined with machine learning to predict the impact of urban structure patterns on thermal environment. Urban Clim. 52:101731. https://doi.org/10.1016/j.uclim.2023.101731

Liu C, Lu S, Tian J, Yin L, Wang L, Zheng W (2024) Research overview on urban heat islands driven by computational intelligence. Land 13:2176. https://doi.org/10.3390/land13122176

Meng F, Yan S, Tian G, Wang Y (2023) Surface urban heat island effect and its spatiotemporal dynamics in metropolitan area: a case study in the Zhengzhou metropolitan area, China. Front Environ Sci 11. https://doi.org/10.3389/fenvs.2023.1247046

Mohamed OYA, Zahidi I (2024) Artificial intelligence for predicting urban heat island effect and optimising land use/land cover for mitigation: prospects and recent advancements. Urban Clim 55:101976. https://doi.org/10.1016/j.uclim.2024.101976

Oh JW, Ngarambe J, Duhirwe PN, Yun GY, Santamouris M (2020) Using deep-learning to forecast the magnitude and characteristics of urban heat island in Seoul Korea. Sci Rep 10:3559. https://doi.org/10.1038/s41598-020-60632-z

Phelan PE, Kaloush K, Miner M, Golden J, Phelan B, Iii HS, Taylor RA (2015) Urban heat island: mechanisms, implications, and possible remedies. Annu Rev Environ Resour 40:285–307. https://doi.org/10.1146/annurev-environ-102014-021155

Ruefenacht L, Acero JA (2017) Strategies for Cooling Singapore: a catalogue of 80+ measures to mitigate urban heat island and improve outdoor thermal comfort. ETH Zurich, Switzerland. https://doi.org/10.3929/ethz-b-000258216

Rupard M (2019) Urban heat islands: causes, impacts, & mitigation

Shamsaei M, Carter A, Vaillancourt M (2024) Using construction and demolition waste materials to alleviate the negative effect of pavements on the urban heat island: a laboratory, field, and numerical study. Case Stud Constr Mater 20:e03346. https://doi.org/10.1016/j.cscm.2024.e03346

Sheng T, Zhang Z, Qian Z, Ma P, Xie W, Zeng Y, Zhang K, Sun Z, Yu J, Chen M (2025) Examining urban agglomeration heat island with explainable AI: an enhanced consideration of anthropogenic heat emissions. Urban Clim 59:102251. https://doi.org/10.1016/j.uclim.2024.102251x

Street Transportation Cool Corridors Program [WWW Document] (n.d.) https://www.phoenix.gov:443/streets/coolcorridors. Accessed on 14 Mar 2025

Tanoori G, Soltani A, Modiri A (2024) Machine learning for urban heat Island (UHI) analysis: predicting land surface temperature (LST) in urban environments. Urban Clim 55:101962. https://doi.org/10.1016/j.uclim.2024.101962

Tsokov S, Lazarova M, Aleksieva-Petrova A (2022) A hybrid spatiotemporal deep model based on CNN and LSTM for air pollution prediction. Sustainability 14:5104. https://doi.org/10.3390/su14095104

US EPA (n.d.) Heat island impacts [WWW Document]. https://19january2017snapshot.epa.gov/heat-islands/heat-island-impacts. Accessed on 13 Mar 2025

Yow DM (2007) Urban heat islands: observations, impacts, and adaptation. Geogr Compass 1:1227–1251. https://doi.org/10.1111/j.1749-8198.2007.00063.x

Yuan T (2024) The role of green infrastructure in mitigating the urban heat island effect. Open J Appl Sci 14:3155–3164. https://doi.org/10.4236/ojapps.2024.1411207

Zuccarini E (2024) Urban heat Island. Machine learning models for analysis and maker approach for mitigation. In: Proceedings of the international workshop on artificial intelligence for climate change, Italian workshop on planning and scheduling, RCRA workshop on experimental evaluation of algorithms for solving problems with combinatorial explosion, and SPIRIT workshop on strategies, prediction, interaction, and reasoning in Italy (AI4CC-IPS-RCRA-SPIRIT 2024) co-located with 23rd international conference of the Italian association for artificial intelligence AIxIA 2024. Presented at the AI4CC-IPS-RCRA-SPIRIT 2024, Italy

Policy Implications of Integrating Remote Sensing Data into Urban Policy and Planning

Rahul Mishra, Farhan Khan, and Bhumika Das

Abstract As Indian cities grow and change rapidly, the need for smarter, more responsive urban planning has never been greater. This paper looks at how geospatial technologies—like satellite imagery, remote sensing, and GIS—can play a powerful role in helping cities become more liveable, sustainable, and better prepared for the future. While India has made progress in adopting digital tools, many challenges remain. Urban data is often scattered across different government departments, legal rules for data sharing are unclear, and local authorities frequently lack the training or tools to make the most of geospatial systems. To understand how things can improve, this study draws insights from both Indian policy documents and successful examples from other parts of the world, like Singapore's Smart Nation strategy and Barcelona's climate-focused city planning. These global cases show what's possible when technology, good governance, and community involvement come together. The paper offers practical suggestions tailored to India's context, such as updating urban master plans with satellite data, encouraging public–private partnerships, building open data platforms for city-level information, and strengthening data privacy laws. It also stresses the need to train city officials and planners to use these tools effectively. Ultimately, the research highlights that geospatial technologies shouldn't just be seen as technical upgrades; there are tools for making urban development more inclusive, transparent, and forward-thinking. With the right policies and a commitment to collaboration, India can use these tools to build cities that work better for everyone.

Keywords Urban policy · Urban planning · Urban heat islands · Land surface temperatures · Urban sprawl

R. Mishra · F. Khan (✉)
Rungta College of Engineering and Technology, Bhilai, India
e-mail: farhan1@rungta.ac.in

Rungta International Skills University, Bhilai, India

B. Das
Department of Civil Engineering, Lakhmi Chand Institute of Technology, Bilaspur, Chhattisgarh, India
e-mail: dr.bhumika.das@lcit.edu.in

B. Kozhikkodan Veettil and B. Das (eds.), *Remote Sensing of Urban Heat Islands*, Cities and Nature, https://doi.org/10.1007/978-3-032-11580-5_10

1 Introduction

1.1 The Urbanization Challenge

Urban areas are growing at an unprecedented rate, concentrating people and assets in expanding cities. This creates multiple challenges: dense population in hazard-prone zones, environmental degradation, and strained infrastructure. For example, coastal urban areas occupy only about 2% of the Earth's land but host roughly 10% of global (and 13% of urban) population, making them highly vulnerable to climate-driven disasters (Aziz et al. 2024). Rising extreme events—heavy rainfall, cyclones, storm surges—are exacerbating urban flooding in these areas. In this context, integrating new data streams is essential to manage rapid urbanization sustainably. Remote sensing can provide the high-resolution, up-to-date information needed to understand and respond to these trends.

Rapid urbanization presents profound challenges globally. Cities now house a majority of people, and this share is expected to grow. About 47% of the world population lived in urban areas in 2000, rising to roughly 70% by 2050. This surge puts pressure on infrastructure, environment, housing, and governance (Perveen et al. 2017). In India alone, Class I cities (over 100,000 population) grew from 289 in 1991 to 433 by 2001, with megacities multiplying. Policymakers must manage rapid sprawl, environmental degradation, and resource constraints under time pressure. Addressing these challenges requires new tools and data sources to monitor and guide urban growth.

1.2 Importance of Evidence-Based Planning

Today's urban development can no longer rely solely on outdated methods like occasional surveys or static maps. These traditional tools often fall short when it comes to keeping up with the fast pace of change in our cities. Instead, we're seeing a shift towards evidence-based planning—an approach that draws on accurate, real-time data to make smarter and more responsive decisions.

Thanks to advancements in technology, planners now have access to a wealth of information through remote sensing, big data, and advanced analytics. This digital transformation allows for a deeper understanding of urban environments, helping policymakers anticipate challenges rather than just react to them. For instance, integrating machine learning and big data tools makes it possible to test different scenarios, predict outcomes, and fine-tune plans before they're implemented.

As highlighted by recent research (Son et al. 2023), these innovations are not just add-ons but have become essential components of modern urban governance. By leveraging spatial data and analytical models, urban planners can more effectively manage land use, address infrastructure gaps, and respond to environmental concerns.

Ultimately, evidence-based planning empowers cities to be more agile, inclusive, and resilient in the face of rapid urbanisation.

1.3 What is Remote Sensing and Why Does It Matter

Remote sensing refers to collecting information about the Earth from a distance, typically via satellite or aerial sensors. It generates continuous, systematic data on land cover, infrastructure, climate variables, and more. Such data provides a synoptic view of cities and their environment over time. For example, satellite imagery can track the expansion of built-up areas, loss of green space, or surface temperatures across an entire city. This broad coverage overcomes the limitations of point surveys and incomplete records. Moreover, modern remote sensors (optical, thermal, LiDAR, radar) combined with data analytics can extract detailed insights—from building footprints and road networks to vegetation health and flood extents. Because urban phenomena are spatially complex, integrating remote sensing into planning gives policymakers a powerful evidence base. It allows issues to be visualized on maps, trends to be quantified objectively, and new (otherwise hidden) patterns to emerge.

1.4 Objectives of the Chapter

This chapter sets out to unpack the transformative role that remote sensing technologies can play in shaping more effective, inclusive, and forward-looking urban policy and planning. As cities around the world face mounting pressures from rapid urbanisation, environmental degradation, and infrastructural stress, the need for responsive and data-driven decision-making has become more pressing than ever.

The central objective of this chapter is to examine how integrating remote sensing data can address these challenges, offering new pathways for evidence-based, geospatially aware urban governance. The chapter begins by contextualising the urbanisation challenge, with particular reference to rapidly growing economies such as India, where expanding cities often outpace traditional planning mechanisms. It introduces the concept of evidence-based planning and establishes the importance of integrating dynamic data sources into policy design and urban development strategies. Remote sensing, in this context, emerges as a vital technological tool capable of offering high-resolution, real-time spatial data that can bridge the gap between existing planning frameworks and the dynamic realities of urban growth. To provide a comprehensive understanding, the chapter charts the evolution of remote sensing in urban contexts. It traces its journey from early satellite imagery and static cartography to the present-day use of AI-enhanced remote sensing, IoT-enabled sensors, and real-time data monitoring systems. These advancements have expanded the potential of spatial data far beyond land classification or environmental monitoring, positioning remote sensing as a core component in smart city planning, resource management, and risk assessment. The next section investigates the current urban policy landscape, with a specific focus on India. It explores how urban governance is structured across multiple tiers of government and highlights key policy instruments such as Master Plans, the AMRUT mission, and Smart Cities initiatives. Within this landscape, the chapter identifies critical data gaps pointing to the inadequacies of conventional data collection methods, and the resultant challenges in forecasting, scenario modelling, and spatial planning. It underscores the need for a paradigm shift towards spatial data integration and multi-source analytics to enable agile and informed governance. The chapter also delves into the functional applications of remote sensing in urban planning. This includes monitoring urban sprawl, detecting informal settlements, tracking land-use change, mapping infrastructure networks, and evaluating environmental risks such as flooding or heat islands. These functions directly support major urban missions, such as AMRUT and Smart Cities, by enhancing planning precision, improving service delivery, and bolstering climate resilience.

Furthermore, the chapter critically analyses the institutional and policy-level challenges that hinder the mainstreaming of remote sensing in planning processes. These include fragmented data ownership, technical skill shortages at the urban local body (ULB) level, outdated legal frameworks, and ethical concerns related to data privacy. In response, the chapter outlines the policy implications of these challenges, advocating for reforms in data governance, inter-agency coordination, and capacity building. Drawing from international best practices, such as Singapore's Smart Nation framework, Barcelona's climate-sensitive planning strategies, and the use of LiDAR in North America, the chapter identifies actionable lessons for the Indian context. These examples serve as benchmarks for how cities can effectively institutionalise geospatial technologies to foster sustainable urban growth. Ultimately, the chapter concludes with recommendations tailored for Indian urban policy, suggesting steps like embedding remote sensing into statutory Master Plans, creating open-access urban geospatial portals, promoting public–private innovation ecosystems, and reforming legal frameworks to support ethical and transparent data use. By bringing together theoretical insights, practical case studies, and current policy evaluations, this chapter aims to guide urban planners, technocrats, and policymakers in leveraging remote sensing as a catalyst for better, data-informed urban futures.

2 Evolution of Remote Sensing in Urban Contexts

2.1 Historical Overview of Remote Sensing

The use of remote sensing for urban analysis dates back to the mid-twentieth century. Early applications involved aerial photographs and declassified satellite images (e.g. CORONA) for basic land-use mapping. From the 1970s onward, the launch of sensors like Landsat enabled routine monitoring of cities at coarse resolution.

Over time, spatial resolution has dramatically improved (Sentinel, WorldView, PlanetScope), allowing increasingly fine detail. In parallel, analytical methods matured from manual photointerpretation to GIS-based analysis. By the 1990s, planners were using satellite imagery for census-independent population estimates and to update maps of urban extents. In the 2000s, object-based image analysis and commercial satellite constellations further refined urban land-use mapping. Now, the trajectory is towards real-time and 3D monitoring technologies such as unmanned aerial vehicles (UAVs), mobile mapping (like Google Street View), and very-high-resolution satellites provide near-continuous views of cities.

Urban remote sensing has roots in mid-twentieth-century aerial photography. Early planners used film photos and maps to document city form. The launch of Earth-observing satellites revolutionized this field. Since Landsat 1 in 1972, successive spaceborne sensors have provided systematic land-cover data. Landsat's 80–30 m imagery allowed analysts to classify land use over large areas. In the 1990s–2000s, higher-resolution satellites (e.g. SPOT, IRS, Ikonos) emerged, yielding 10–1 m images. Each advance enabled finer urban mapping: e.g. India's Thematic Mapper (30 m) was first used to map city sprawl, and SPOT (10 m) to update land-use details. Over time, digital sensors and GIS supplanted paper maps, setting the stage for automated analysis and integration with other data (census, surveys, sensor networks).

Remote sensing has a long history, dating back to the mid-nineteenth century when cameras were first made airborne using balloons and kites. The advent of aircraft further enhanced opportunities to capture aerial photographs. During World War I, significant advancements occurred in remote sensing as data and information were acquired using this technique. Over the years, remote sensing has evolved into a key source of environmental data for local to global analysis, representing one of the largest publicly accessible databases about our planet and its cities (Wellmann et al. 2020).

2.2 *Milestones in Urban Remote Sensing Applications*

Key milestones illustrate remote sensing's growing urban role. Early efforts focused on thematic mapping—e.g. updating city boundaries or land-cover classes. More recently, *computer vision* and *machine learning* have been applied to urban imagery. For instance, systematic reviews identify breakthroughs where deep learning models automatically detect buildings from high-resolution images (Son et al. 2023). These algorithms supplant labour-intensive surveys, enabling planners to inventory infrastructure at city scale. Similarly, researchers have shown that panoramic street-view images (from platforms like Google or Baidu) can be analysed with AI to measure urban features such as sidewalk width, tree canopy, and building conditions (Marasinghe et al. 2024). These methods vastly increase urban coverage and reduce human bias. Another milestone is the integration of socio-economic data with remote images: e.g. using satellite nightlights or road networks to infer population distribution and informal settlements. Collectively, these advances have moved remote sensing from a niche mapping tool to a central component of "smart" urban analysis.

Key milestones have shaped urban remote sensing. Early uses included identifying urban boundaries and general land cover. For instance, India's NRSC applied Landsat TM and IRS-1A imagery in the 1990s to inventory land use and support regional plans (e.g. Delhi in 1993). The introduction of IRS-1C/1D (5.8 m) and the Cartosat series (2.5–1 m) around the 2000s-enabled detailed city maps. These high-resolution images were used for revising city development plans (e.g. Hyderabad, Ahmedabad). Currently, the Indian government uses Cartosat data for its National Urban Information System (NUIS), generating multi-scale thematic maps for hundreds of towns. Globally, similar progress occurred: from Landsat's 1972 launch to today's near-real-time constellations (Sentinel, PlanetScope, commercial satellites), spatial resolution and update frequency have dramatically improved. These developments have allowed planners to move from static map snapshots to continuous monitoring of urban change.

2.3 Shift from Static Maps to Dynamic Planning Tools

Historically, planning relied on static maps created from infrequent surveys. Today, remote sensing enables *dynamic* planning support. Urban data can be updated continuously: for example, weekly satellite overpasses detect new construction or vegetation change. This real-time monitoring transforms planning from reactive to proactive. Reviews highlight that advanced data methods can make planning "smarter", leveraging up-to-date spatial inputs (Son et al. 2023). For example, algorithmic models now predict urban expansion trajectories by analysing past imagery and growth patterns. Dynamic dashboards can display live feeds (e.g. thermal maps for heat, flood inundation extents after storms) to inform emergency response and infrastructure management. Essentially, remote sensing has shifted the paradigm from fixed map books to an evolving geospatial intelligence system, improving the timeliness and relevance of urban policies.

Advances in sensor technology have transformed planning from static mapping to dynamic simulation. In the past, urban plans were based on infrequent data collection and paper diagrams. Today, city models and digital twins simulate complex interactions. A notable example is *Virtual Singapore*, a national 3D digital twin of Singapore's entire urban environment. Built from aerial surveys, LiDAR, and street-level scans, this platform enables planners to visualize and test interventions virtually. As the Virtual Singapore team reports, "the 3D city model… acts as a digital twin of the city", allowing planners to experiment with designs (e.g. new bridges or green spaces) without physical trials. Similarly, many cities now use GIS-based dashboards and interactive models. These tools harness satellite imagery along with real-time sensor feeds, moving planning towards continuous, data-driven decision support (see Sects. 4 and 5).

2.4 Emerging Trends: AI, IoT, and Real-Time Monitoring

A convergence of emerging technologies is enhancing urban remote sensing. Artificial intelligence (AI) and machine learning (ML) are enabling extraction of complex patterns from imagery. Systematic reviews report that local governments are early adopters: AI is applied for tasks like optimizing land use, simulating development scenarios, and extracting demographic information (Son et al. 2023). Notably, (a) pilot AI applications are paving the way for wider adoption; (b) big data—including remote sensing archives—is integral to effective AI-driven planning; and (c) combining AI with human expertise (i.e. a "convergence of artificial and human intelligence") is crucial for holistic planning (Son et al. 2023). Geospatial AI (GeoAI) specifically fuses CV and ML, offering strong advantages over traditional methods. It provides wider geographic coverage, minimizes sampling bias, and improves speed and efficiency (Marasinghe et al. 2024). For instance, GeoAI can automatically map green spaces or structural damage city-wide from satellite data, tasks impossible manually.

Concurrently, the Internet of Things (IoT) is providing new data streams. Urban sensors (air quality monitors, traffic counters, smart meters) continuously collect environmental data. Integrating these with remote sensing enables real-time monitoring of conditions. Recent studies highlight trends like IoT-enabled climate monitoring and autonomous systems. For example, IoT sensors combined with satellite data can power digital twins of cities for climate adaptation, and ML-driven systems optimize services such as smart waste collection (Li et al. 2023). In sum, the integration of AI and IoT is making urban planning increasingly data-driven and adaptive. The trajectory points to real-time, automated analysis of remote sensing feeds, forming a foundation for agile, evidence-based policy.

Today's frontiers include artificial intelligence, Internet-of-Things (IoT) sensors, and real-time analytics. AI and machine learning can automatically extract urban features (buildings, roads, heat islands) from imagery. Son et al. (2023) emphasize that big data and AI are integral to future urban planning: successful smart-city implementation will "involve collaboration between stakeholders" and the fusion of AI with human expertise. IoT networks complement satellites by providing hyper-local measurements. For example, Barcelona has deployed citywide sensor grids monitoring air quality, traffic, and waste bins; this network feeds into planning for environmental and mobility improvements. Singapore's Virtual Singapore integrates not only aerial maps but also real-time sensor data from its Smart Nation program. Together, these trends mean that urban remote sensing is no longer offline: data streams (satellite, drone, sensor) are continuously ingested into planning platforms. Policymakers are thus moving from one-off surveys to dynamic monitoring systems that enable rapid policy feedback and adaptation.

3 Current Urban Policy Frameworks and Data Gaps

3.1 *Structure of Urban Governance in India*

Urban governance in India is multi-layered. Cities are typically governed by municipal corporations or councils, under state oversight and guided by national ministries (e.g. Ministry of Housing and Urban Affairs). Planning is also influenced by statutory bodies (development authorities, state urban development boards) and schemes at different levels. For example, the Smart Cities Mission and Atal Mission for Rejuvenation and Urban Transformation (AMRUT) are national programs that allocate funds to local bodies for infrastructure and service improvement. However, decision-making often suffers from fragmented responsibilities and limited inter-agency coordination. This institutional complexity can slow data sharing. Urban planners rely on mandated instruments (master plans, zonal plans, urban transport plans) but these often lack the flexibility to incorporate new types of data quickly. In many cases, even though remote sensing data are available, there is no streamlined mechanism to integrate them into local government databases or statutory planning documents. Strengthening spatial data governance is thus a policy priority.

India's urban governance is multi-layered, involving central ministries, state governments, urban local bodies (ULBs), and parastatal agencies. The Ministry of Housing and Urban Affairs (MoHUA) oversees national programs, while states implement policies through municipal corporations and development authorities. However, NITI Aayog (2021) notes systemic challenges: data fragmentation across agencies, outdated development regulations, and limited technical capacity at local levels hinder effective planning. Many decisions still rely on manual processes. Moreover, urban governance often lacks standardized spatial data sharing: cadastral maps, satellite-derived inventories, and field surveys are maintained separately, if at all. This fragmentation means that even within the same city, planning departments may operate with inconsistent maps and statistics. Addressing these governance gaps requires integrating spatial data into official planning workflows, as recent policy initiatives suggest (see Sect. 3.2).

3.2 *Key Urban Planning Instruments and Guidelines*

India has several formal planning instruments: city development plans, master plans, and local area plans. There are also guidelines for sectoral aspects like building by-laws, land-use zoning, and environmental clearances. Missions like Smart Cities require city-level plans (city development plans) with emphasis on technology and infrastructure. Under AMRUT, cities prepare Service-Level Improvement Plans. While these frameworks encourage evidence-based approaches on paper, in practice many continue to use conventional data sources: census, economic surveys, or limited field studies. Spatial data are explicitly mentioned in some guidelines (e.g.

"GIS-based Master Plans" in the Model Building Regulations), but actual use is uneven. Urban planning standards have not fully caught up with the pace of remote sensing innovation. In effect, many policies acknowledge the value of geospatial data but lack clear protocols for regularly updating plans with remote-sensed information.

India has instituted several flagship programs that implicitly encourage geospatial approaches. The *Smart Cities Mission* (2015) covers 100 cities and explicitly calls for "smart solutions" underpinned by digital infrastructure. Under this program, cities are developing multi-layer GIS databases ("40 layer GIS mapping") and setting up Integrated Command and Control Centres (ICCCs) to pool data from diverse sources. For example, a national *Geospatial Management Information System* aggregates project data across cities, and the India Urban Observatory integrates multi-parameter data (IoT sensors, satellite, crowdsourced) into an interactive dashboard. Similarly, the *Atal Mission for Rejuvenation and Urban Transformation* (AMRUT 2015) focuses on water, sewage, transport, and green spaces, with a GIS-based master planning sub-scheme. AMRUT requires the creation of detailed urban geodatabases (1:4000 scale) and has piloted GIS-driven local area planning. Furthermore, draft policies on geospatial data (e.g. the National Geospatial Policy) are easing data restrictions. Together, these instruments signal a shift: India's urban programs are increasingly built around spatial data frameworks, even if implementation is still uneven.

3.3 *Limitations of Traditional Data Sources*

Traditional urban data collection faces well-known limitations. Census data, land records, and household surveys are often outdated (e.g. India's decadal census) or infrequently refreshed. They also tend to be coarse at the ward or city level and can omit informal settlements or recent land-use changes. Physical surveys are costly and time-consuming, limiting coverage. As a result, planners may lack timely insight into rapidly evolving phenomena such as seasonal population shifts, new construction, or emerging risk zones. Moreover, non-spatial data (statistics) often miss locational context. This creates blind spots: planners may be unaware of, say, the spatial pattern of heat stress or flood risk at a neighbourhood scale. Remote sensing helps fill these gaps by providing continuous, high-resolution spatial data. Still, a challenge is integrating these new sources with legacy data systems. Advanced analytical tools (including AI) can merge remote-sensed data with surveys but require capacity building.

Despite these policies, conventional data remain problematic. Census and official surveys occur infrequently (every 10 years or less), making them stale for fast-growing cities. Moreover, surveys often lack granularity or consistency: different areas of a city may rely on maps from different years. The NRSC notes that mixed-age survey maps make plan preparation slow, since reconciling data from different years is difficult. In practice, planners frequently encounter mismatched or missing data when drafting plans. The result is that urban plans are often based on the "best available" yet outdated information, undermining their accuracy. These gaps underscore the need for alternative sources of spatial data that can be updated more frequently. Remote sensing imagery, which can be acquired at regular intervals, offers one solution to bridge this gap.

3.4 Need for Spatial Data Integration

To address these gaps, policy must prioritize spatial data integration. Modern GIS and remote sensing tools enable overlaying diverse datasets—land use, population density, infrastructure networks, environmental indicators—in a single platform. For instance, GIS-based zoning and spatial analysis have dramatically improved land-use optimization (Li et al. 2023). These techniques allow planners to identify underserved areas, optimize resource allocation, and simulate development scenarios. The planning instruments should mandate the use of up-to-date satellite imagery and spatial analysis. This means developing standards for incorporating remote sensing into official maps and reports, and training officials to use geospatial decision support systems. Reviews emphasize that (resource-constrained) cities benefit greatly from integrating GIS in planning: it makes land-use planning more efficient and sustainable (Li et al. 2023). Ultimately, integrating remote-sensed spatial data can overcome the inertia of traditional data sources, leading to more responsive urban governance.

Given the shortcomings above, there is a growing consensus on the need for integrated spatial datasets. Urban problems are inherently geographic, yet many planning decisions today ignore spatial context due to siloed data. Remote sensing and GIS together can integrate diverse layers—land use, infrastructure, demographics—in one platform. As the Indian government report states, coupling remote sensing with GIS "facilitates planners in making decisions" by providing reliable spatial data quickly to officials and the public. Such integration would enable more dynamic policy scenarios (e.g. what-if analyses) and better monitoring of outcomes. In short, mainstreaming spatial data—satellite imagery, aerial LiDAR, sensor networks—within city governance is key to closing the evidence gap. Overcoming technical and institutional barriers to this integration is therefore a central policy challenge in India's urban agenda.

4 Functional Role of Remote Sensing in Urban Planning

4.1 Monitoring Urban Sprawl and Land-Use Change

One core function is tracking urban expansion and land-cover change. Satellite imagery can map built-up areas, vegetation, water bodies, and impervious surfaces over time. For example, consistent use of GIS and remote sensing tools enables cities to efficiently optimize land use and curb unregulated sprawl (Li et al. 2023). Such spatial analysis reveals where cities are encroaching on agricultural land or forests and helps enforce zoning regulations. Continuous monitoring can alert planners to unexpected development or encroachments. Remote sensing also supports predictive modelling: combining past urban footprints with socio-economic drivers yields forecasts of likely growth areas. This informs infrastructure planning (e.g. where to extend roads or utilities). In short, by providing objective maps of land use and its changes, remote sensing underpins evidence-based management of urban growth.

Remote sensing excels at charting urban expansion. Multitemporal satellite images can quantify how quickly cities sprawl into surrounding land. For example, analysis of Landsat imagery shows the footprint of cities growing year by year. At fine scales, very-high-resolution data allow mapping individual buildings and roads. In practice, planners use these maps to update land-use inventories and enforce zoning. The USGS notes that satellites help "track the growth of a city and changes in farmland or forests over several years". Indian agencies have applied this: RS-based maps identified informal settlements and peri-urban growth outside planned areas, prompting targeted infrastructure deployment. Thus, remote sensing provides an objective view of where and how fast cities expand, informing policies on containment, densification, or provision of services.

Remote sensing enables high-resolution tracking of urban sprawl, agricultural land loss, and informal settlement growth. For example, the Normalized Difference Vegetation Index (NDVI) helps detect green space shrinkage in cities, informing strategies to preserve urban greenery (Wulder et al. 2019).

4.2 Environmental Risk and Disaster Preparedness

Remote sensing plays a critical role in identifying and managing environmental risks. Urban Flooding: Planners use satellite and aerial imagery to delineate flood-prone zones, map drainage patterns, and simulate inundation. Coastal cities, despite small geographic extent, face outsized flood risk as climate change intensifies storms (Aziz et al. 2024). Remote sensing data (e.g. digital elevation models, surface reflectance) feed hydrological models to improve flood risk maps. They also enable post-event damage assessment (satellite imagery before/after a flood) to guide relief and adaptation efforts.

Urban planners must also manage environmental risks, a task greatly aided by remote sensing. Satellites routinely monitor natural hazards and environmental quality. For instance, thermal imagery and multispectral sensors can detect urban heat islands and air pollution patterns. Satellites track floods and wildfires, giving authorities early warning over large areas. In high-resolution use cases, LiDAR data have improved flood modelling: one Toronto study found that a 1 m LiDAR digital elevation model captured curbs and streets, refining urban hydrology simulations and stormwater planning. LiDAR 3D mapping also helps assess solar potential (by roof slope/orientation) and emergency response times in high-rise zones. Integrating such data strengthens disaster resilience. For example, real-time satellite imagery (e.g. from Sentinel) can identify inundated areas during a flood, supporting evacuation and infrastructure planning. In sum, remote sensing contributes critical spatial context to environmental planning—enabling risk mapping, monitoring of ecosystem services, and evidence-backed mitigation strategies.

Satellite data can forecast flood-prone zones, heat islands, and landslide risks. In Chennai, for instance, remote sensing was pivotal in mapping flood plains post-2015 floods, aiding in disaster preparedness and mitigation efforts (Wellmann et al. 2020).

Heat and Climate Vulnerability: Urban heat islands and heat waves pose growing threats. Recent studies show the importance of real-time monitoring: "Remote sensing can offer valuable data to help urban planners and policymakers develop effective strategies for enhancing climate resilience". For instance, satellite thermal data identify heat hotspots within cities (e.g. dense built areas lacking green cover). Spatial overlay with demographics highlights vulnerable communities. Scientometrics reviews note that the use of satellite remote sensing in heat-risk studies has surged over the past decade (Cheval et al. 2024). This trend reflects planners' need for high-resolution temperature data to design cooling interventions (e.g. parks, cool roofs). Integrated approaches combining demographic, land-use, and thermal data help target heat mitigation.

Industrial and Pollution Hazards: Remote sensing also detects industrial "hotspots". A case study in Indonesia used multi-sensor data (Landsat surface temperature, NDVI, emission inventories) with machine learning to map industrial heat islands and carbon emissions (Suprijanto et al. 2025). They found industrial zones raising local temperatures up to 1500 m from the plant and destroying over 80% of nearby vegetation. Such insights are invaluable for zoning regulations and mitigation (e.g. buffer zones, clean energy retrofits). Air quality sensors and aerosol imagery can similarly identify pollution plumes from factories or traffic corridors. By quantifying environmental impacts spatially, remote sensing informs risk mitigation (e.g. emergency planning, greenbelt policies).

Ecosystem Disservices: Urban green spaces provide benefits, but can also have "disservices" (pollen, pests) that affect health. Planners must balance services and disservices. Recent reviews emphasize that ignoring urban disservices exacerbates problems (Veibiakkim et al. 2025). Remote sensing of vegetation health and land cover can flag areas where, for example, dense vegetation correlates with allergic pollen levels. Integrating this with social data can lead to planting guidelines or the removal of invasive species. Overall, combining remote environmental monitoring with social vulnerability analysis helps cities prepare for a range of environmental risks.

4.3 Infrastructure Mapping and Utility Planning

Remote sensing enables detailed mapping of physical infrastructure at city scale. High-resolution imagery and LiDAR data can identify roads, rail lines, buildings, and utility layouts. For example, advanced computer-vision algorithms have been developed to automatically extract building footprints from aerial images (Son et al. 2023). These techniques allow planners to maintain up-to-date inventories of built-up areas without exhaustive field surveys. Utility networks (water pipes, power lines) are not directly visible from satellites, but planners can infer infrastructure density from patterns of development (e.g. street network extracted from imagery). Remote sensing also aids urban retrofitting: identifying derelict areas for redevelopment, mapping solar-potential rooftops, or locating suitable corridors for new transit lines. Moreover, integration with GIS lets planners perform spatial analyses (e.g. service area mapping for hospitals or utilities). By reducing the cost and time of infrastructure surveys, remote sensing supports more proactive utility planning and "smart" infrastructure investment.

High-resolution geospatial data are invaluable for infrastructure development. Satellite and aerial imagery can map transportation networks, utility corridors, and urban amenities across entire cities. The NRSC notes that remote sensing has been widely used for telecommunication and transport network planning, highway development, and accessibility analysis. For example, planners can derive road network maps to plan new routes or mass transit lines. Pipeline and sewer routes can be optimized using up-to-date terrain models. Satellite data also help inventory green spaces, water bodies, and civic facilities, ensuring infrastructure roll-out matches actual ground conditions. Furthermore, combining imagery with GIS enables predictive maintenance: detecting landslide-prone slopes or subsidence can guide priority in water and power networks. In India, such practices are beginning to emerge under national programs: Smart Cities dashboards, for instance, compile map layers of utilities (water lines, streetlights, public toilets) to coordinate upgrades. Overall, remote sensing supplies the spatial backbone that lets planners design and retrofit city infrastructure efficiently.

4.4 Supporting Smart City and AMRUT Missions

Remote sensing data align closely with India's Smart Cities Mission and AMRUT objectives. These programs emphasize digital solutions and performance-based planning. For instance, smart-city dashboards often integrate satellite imagery layers (green cover, built-up density, water bodies) to monitor urban indicators. Data-driven applications are emerging: AI-assisted systems optimize waste routes or traffic signals based on sensor and satellite data (Li et al. 2023). Remote sensing also provides the foundation for urban digital twins—real-time virtual models of the city combining sensor feeds and spatial data. Such platforms help test policy interventions (e.g. how planting trees might reduce heat). Importantly, remote-sensed evidence can democratize planning: open-access maps and apps allow citizens to engage in monitoring (e.g. spotting illegal dumping or encroachment).

For AMRUT's focus on service delivery, remote sensing helps track green space and water bodies (key performance indicators) accurately and regularly. Satellite data can verify whether tree-planting targets are met or water coverage restored. In sum, integrating remote sensing into these initiatives makes them more accountable and effective. Reviews stress that the convergence of AI, IoT, and spatial analytics is creating "actionable insights" for urban policy (Li et al. 2023). To capitalize on this, policymakers should explicitly incorporate remote sensing into smart city guidelines and investment criteria. In doing so, urban governance can truly become spatially savvy, evidence-based, and responsive to twenty-first-century challenges.

National initiatives like the Smart Cities Mission and AMRUT explicitly call for data-driven planning, where remote sensing can play a supportive role. Under Smart Cities, Integrated Command and Control Centres aggregate data across domains (traffic, energy, safety), and could integrate satellite-derived layers (e.g. land use, green cover) into their dashboards. Many Smart Cities are already adopting GIS platforms to monitor urban performance. Similarly, AMRUT's emphasis on water, sewage, and transport can be strengthened by spatial analysis: for example, satellite data can identify informal water pipelines or drainage bottlenecks. Indeed, AMRUT launched a GIS-based master plan scheme for hundreds of cities, explicitly building geodatabases for urban services. Looking ahead, policy guidelines might mandate the use of remote sensing data as a planning input. As these missions evolve, embedding spatial data standards and analytics into project appraisal could ensure that investments align with actual needs. In short, current policy frameworks are opening the door to broader use of geospatial tools, and realizing this promise will require active promotion of remote sensing as a planning resource.

5 Policy and Institutional Challenges

5.1 Fragmentation Across Government Departments

India's geospatial data landscape suffers significantly due to fragmentation across multiple government departments and agencies. Each department—ranging from agriculture, urban development, and environment to transportation and disaster management—collects and uses spatial data independently, often based on their specific mandates and operational needs. This compartmentalized approach results in overlapping responsibilities, duplication of efforts, and a lack of interoperability among datasets. Departments might use different coordinate systems, mapping scales, data collection methodologies, and storage formats, which severely hampers the integration of data across different domains. As a result, the potential of geospatial data to offer comprehensive, multi-sectoral insights remains largely untapped (Ahluwalia 2019).

This fragmentation is further exacerbated by the absence of a unified geospatial governance structure or a central coordinating body that enforces common standards and protocols. While the National Spatial Data Infrastructure (NSDI) was envisioned to provide a shared framework and facilitate inter-agency data sharing, its execution has been piecemeal and slow (Technology 2021b). Many departments are reluctant to share their data due to bureaucratic inertia, jurisdictional rigidity, or even concerns over data sensitivity and ownership. Consequently, datasets often remain underutilized or inaccessible to other government bodies, academic researchers, private sector players, and civil society organizations.

The lack of a centralized repository or open data policy adds another layer of inefficiency. Currently, users must approach different agencies separately to access data, which is time-consuming and may involve administrative hurdles. Moreover, there is often little transparency in how data is collected, maintained, or updated, which affects its accuracy and reliability (Group 2021).

The fragmentation also negatively impacts national programs that depend on spatial data integration, such as Smart Cities, Digital India, and urban planning initiatives under AMRUT. A more collaborative and standardized approach would yield far greater outcomes in such programs, enabling data-driven policy decisions and real-time monitoring of developmental goals (Aayog 2021).

To address these systemic issues, a comprehensive National Geospatial Policy is urgently required. This policy should clearly define roles and responsibilities across departments, mandate interoperability standards, and encourage data sharing by default unless there are compelling reasons for confidentiality. The establishment of a central geospatial authority or a cross-sectoral coordinating body under the Prime Minister's Office or NITI Aayog could provide the necessary oversight and coordination.

Furthermore, leveraging emerging technologies such as cloud computing, blockchain for data validation, and APIs for seamless data exchange can improve the technical robustness of inter-departmental cooperation. Government initiatives like the India Geo-Platform (ISRO 2023), developed by ISRO, and other data integration tools should be strengthened and aligned with global standards.

Public–private partnerships (PPPs) also offer immense potential in this regard. Private players, especially startups in the geospatial domain, can bring in innovation and agility, while the government can ensure accountability and data sovereignty. Collaborative models would ensure that high-quality, real-time spatial data can be generated, maintained, and utilized for national development.

In conclusion, overcoming fragmentation in geospatial data management requires institutional reform, technological upgrades, and a cultural shift towards openness and collaboration. Only then can India harness the full potential of geospatial technologies to achieve its socio-economic and environmental objectives.

5.2 *Lack of Technical Capacity in Urban Local Bodies (ULBs)*

Urban local bodies (ULBs) play a critical role in managing and planning India's rapidly urbanizing landscape. From preparing city master plans to managing essential services like water supply, sanitation, and traffic, ULBs are at the heart of local governance. Despite this centrality, many ULBs in India lack the technical capacity required to effectively leverage geospatial data and technologies. This limitation becomes a significant barrier when integrating Geographic Information Systems (GIS), remote sensing, and spatial analytics into urban governance and planning processes.

One of the primary challenges lies in the shortage of skilled personnel. Most ULBs do not have dedicated GIS professionals or technical teams capable of collecting, managing, and analysing spatial data. Even in cases where some expertise is available, it is often inadequate or outdated, leading to ineffective utilization of existing geospatial platforms. Additionally, frequent staff transfers and the lack of institutional memory make it difficult to sustain capacity-building efforts (NIUA 2023).

Infrastructure constraints are equally concerning. Many ULBs lack the necessary hardware, software, and high-speed internet connectivity to operate modern geospatial platforms. The available tools are often legacy systems that do not support interoperability or real-time data visualization. Consequently, ULBs end up relying on traditional methods like paper-based maps or static reports, which are not suited for dynamic urban environments.

Financial constraints further compound these issues. Geospatial technology requires investment—not just in software and data acquisition but also in training and long-term maintenance. However, budgetary allocations for such technological upgrades are often minimal in ULBs, as funds are directed towards more immediate infrastructure needs. Even when funding is available through centrally sponsored schemes like Smart Cities or AMRUT, many smaller ULBs struggle to access or effectively utilize these funds due to procedural complexities or a lack of proposal-writing expertise (India 2015).

To bridge this gap, structured capacity-building programs must be introduced across all tiers of urban governance. These programs should include hands-on training modules in GIS, data management, and spatial analytics tailored to the practical needs of ULBs. Partnerships with technical institutions like the Indian Institute of Remote Sensing (IIRS), National Institute of Urban Affairs (NIUA), and private sector GIS firms could be established to offer such training in a sustainable, modular fashion.

Moreover, centralized support mechanisms such as shared GIS service centres at the state or regional level could help smaller ULBs that lack individual capacity. These centres can provide on-demand geospatial services, technical guidance, and data processing support, enabling even the resource-constrained ULBs to make data-driven decisions.

Digital governance initiatives also present an opportunity. If properly integrated, platforms like e-Nagar and the Urban Observatory can serve as backbones for spatial data sharing and analysis at the municipal level. The government should work towards embedding geospatial functions into these digital portals to encourage their widespread use.

Finally, a performance-linked incentive model could be introduced, rewarding ULBs that adopt and effectively use geospatial technologies. This would create a healthy competitive environment and motivate local administrations to prioritize technical upskilling.

In summary, strengthening the technical capacity of ULBs is not just about technology; it is about empowering local institutions to make informed, timely, and sustainable decisions. With the right support mechanisms in place, geospatial technologies can transform how Indian cities grow and function.

5.3 Legal and Regulatory Constraints on Geospatial Data

India's regulatory landscape concerning geospatial data has undergone significant transformation over the past few decades. However, despite recent liberalization efforts, there remain numerous legal and regulatory constraints that hamper the efficient collection, sharing, and use of geospatial information. These constraints stem primarily from national security concerns, historical policies, bureaucratic inertia, and an evolving understanding of digital sovereignty and data governance.

Traditionally, geospatial data in India has been tightly controlled by government agencies such as the Survey of India (SoI), Defence Ministry, and Indian Space Research Organisation (ISRO). The logic behind this centralized control was primarily to ensure national security and prevent the misuse of sensitive mapping information. This cautious approach was institutionalized through laws like the Official Secrets Act (1923) and restrictions under the Remote Sensing Data Policy (RSDP) of 2011, which imposed stringent limitations on the access, use, and dissemination of high-resolution spatial data. These restrictions created a bottleneck for both public and private stakeholders interested in leveraging geospatial tools for urban planning, disaster management, environmental monitoring, and infrastructure development. The lengthy approvals, licensing procedures, and lack of clarity regarding data ownership discouraged innovation and foreign investment, particularly in sectors such as real estate, transportation, logistics, and telecommunications.

Recognizing these challenges, the Government of India made a landmark policy shift in 2021 by introducing the Guidelines for Acquiring and Producing Geospatial Data and Geospatial Data Services, issued by the Department of Science and Technology (DST). These guidelines marked a departure from a permission-based model to a self-certification approach for Indian entities, allowing them to collect, process, and use most geospatial data without prior government approval. The move was welcomed as a step towards democratizing spatial data and fostering innovation in sectors like AgriTech, FinTech, and Smart Cities (Technology 2021a).

Despite these progressive changes, several regulatory ambiguities persist. First, the guidelines do not clearly define mechanisms for enforcement, grievance redressal, or quality assurance. Second, there is limited coordination between the DST guidelines and other data-related regulations such as the Information Technology Act (2000) or the new Digital Personal Data Protection Act (2023). This lack of harmonization can lead to confusion and overlapping jurisdictions, particularly in matters involving sensitive or personal location data.

Additionally, the legal framework still lacks clear provisions related to data licensing, intellectual property rights, and liability for data misuse. This uncertainty discourages startups and private sector players from investing in large-scale geospatial projects. Without clarity on who owns the data, who can monetize it, or how derivative works are regulated, the market remains fragmented and underdeveloped.

To unlock the full potential of geospatial technologies, India needs a comprehensive and updated Geospatial Data Act. This law should consolidate existing policies, address security concerns, encourage data openness, and provide legal clarity for public–private partnerships. It should also establish a regulatory body—similar to TRAI or SEBI—for geospatial data governance, with the authority to oversee compliance, standardization, and ethical usage.

In conclusion, while India has taken commendable steps to liberalize its geospatial policy framework, legal and regulatory constraints continue to pose significant hurdles. A forward-looking, transparent, and harmonized legal regime is crucial for building a resilient and innovation-friendly geospatial ecosystem.

5.4 Issues of Data Privacy and Ethical Use

With the proliferation of location-based services, satellite imaging, and GIS technologies, concerns around data privacy and ethical use have become increasingly significant. Geospatial data, especially when combined with personal identifiers or sensitive attributes, can pose serious privacy risks. For instance, real-time mapping of individuals' movement, health records linked with location data, or the identification of vulnerable populations can lead to surveillance or discrimination if misused.

India currently lacks a robust data protection framework specifically tailored for geospatial data. While the Digital Personal Data Protection Act (2023) provides a general framework, it does not address the unique risks posed by spatial datasets. For example, anonymization techniques common in other forms of data may be insufficient in geospatial contexts where location patterns can easily re-identify individuals or communities.

Moreover, the ethical use of geospatial data in policymaking is often overlooked. Projects involving forced relocations, infrastructure development in ecologically sensitive areas, or mapping of slums can have serious ethical implications if not managed carefully. The absence of clear ethical guidelines or institutional review mechanisms for geospatial projects exacerbates the risk.

To safeguard privacy and ensure ethical use, India must establish sector-specific guidelines that govern the collection, sharing, and use of geospatial data. These should include consent mechanisms, redressal systems, and stringent norms for anonymization. Ethics committees or data governance boards should be institutionalized in public agencies handling large-scale geospatial projects.

Further, awareness campaigns and training for government officials, private actors, and researchers on ethical data handling practices can build a more responsible ecosystem. Promoting ethical frameworks like "Privacy by Design" in spatial technologies will also help align innovation with public interest.

6 Policy Implications and the Need for Reform

6.1 Rethinking Urban Policy Through a Geospatial Lens

India's rapid urbanization presents both opportunities and challenges. Traditional urban policies often rely on outdated data and fragmented planning approaches, leading to inefficient resource allocation and inadequate infrastructure development. Integrating geospatial technologies into urban policymaking can revolutionize this process by providing real-time, location-specific data that enhances decision-making.

Geospatial tools enable urban planners to visualize complex spatial relationships, monitor land-use changes, and assess environmental impacts. For instance, satellite imagery and GIS can help identify urban heat islands, track deforestation, and monitor water bodies, facilitating proactive measures to mitigate environmental degradation. Moreover, these technologies support disaster management by mapping vulnerable areas and planning evacuation routes.

The National Geospatial Policy (2022) underscores the importance of geospatial data in urban planning, aiming to make India a global leader in this domain. By promoting the use of geospatial technologies, the policy seeks to enhance transparency, efficiency, and sustainability in urban development (Communications 2023).

However, challenges persist. Data silos, lack of standardized protocols, and limited technical expertise hinder the effective integration of geospatial data into policymaking. To overcome these barriers, it's essential to foster inter-agency collaboration, invest in capacity building, and develop standardized data-sharing frameworks. Additionally, engaging stakeholders, including local communities, in the planning process ensures that urban policies are inclusive and responsive to ground realities (National Geospatial Policy 2022).

6.2 Importance of Open-Access Data and Transparency

Open-access geospatial data is pivotal for fostering transparency, accountability, and innovation in governance. By making data readily available, governments empower citizens, researchers, and businesses to participate actively in decision-making processes, leading to more informed and equitable outcomes.

The National Data Sharing and Accessibility Policy (NDSAP) of 2012 marked a significant step towards democratizing data in India. It emphasized the importance of sharing non-sensitive data generated using public funds, promoting transparency and facilitating research and innovation. Building on this, the National Geospatial Policy (2022) aims to further liberalize access to geospatial data, encouraging private sector participation and fostering a dynamic geospatial ecosystem (National Data Sharing and Accessibility Policy 2012).

Open data initiatives have demonstrated tangible benefits. For example, accessible geospatial data has enabled startups to develop location-based services, improved disaster response through real-time mapping, and enhanced urban planning by providing insights into land-use patterns. Moreover, transparency in data fosters trust between citizens and the government, as policies and decisions become more evidence-based and accountable (Forum 2015).

However, challenges remain. Data quality, standardization, and interoperability issues can limit the utility of open data. Additionally, concerns about data privacy and security necessitate robust frameworks to protect sensitive information. To maximize the benefits of open-access data, it's crucial to invest in data infrastructure, establish clear guidelines for data sharing, and promote data literacy among stakeholders.

6.3 Inter-Agency Collaboration and Governance Models

Effective integration of geospatial data into policymaking requires robust inter-agency collaboration and cohesive governance models. In India, multiple agencies collect and manage geospatial data, often leading to duplication of efforts, inconsistencies, and inefficiencies. Establishing a coordinated framework can streamline data sharing, enhance decision-making, and optimize resource utilization (Contributors 2025).

The National Spatial Data Infrastructure (NSDI) initiative aims to facilitate the sharing and integration of geospatial data across various government departments. By promoting standardized data formats and protocols, NSDI seeks to create a unified geospatial data repository accessible to all stakeholders (Technology 2021b).

Successful inter-agency collaboration requires clear delineation of roles and responsibilities, standardized data collection methodologies, and interoperable systems. For instance, integrating data from the Ministry of Environment with that from the Ministry of Urban Development can provide comprehensive insights into sustainable urban planning. Moreover, leveraging public–private partnerships can bring in technological expertise and innovation, enhancing the overall geospatial ecosystem.

Governance models should also emphasize capacity building, ensuring that agencies have the necessary skills and resources to manage and utilize geospatial data effectively. Regular training programs, workshops, and knowledge-sharing platforms can foster a culture of collaboration and continuous learning. Additionally, establishing oversight bodies can monitor data quality, compliance, and the effectiveness of collaborative efforts.

6.4 Investing in Capacity Building and Training

The successful integration of geospatial technologies into policy-making hinges on the availability of skilled professionals and robust institutional capacities. Investing in capacity building and training ensures that stakeholders can effectively collect, analyse, and utilize geospatial data, leading to more informed and efficient decision-making.

India has recognized the importance of capacity building in this domain. Institutions like the Indian Institute of Remote Sensing (IIRS) and the Indian Institute of Space Science and Technology (IIST) offer specialized courses and training programs in geospatial technologies. Additionally, initiatives like the National Geospatial Programme focus on enhancing the skills of professionals across various sectors (Technology 2021a).

Capacity building should encompass a broad spectrum of stakeholders, including government officials, urban planners, researchers, and community leaders. Tailored training programs can address specific needs, such as urban planning, disaster management, or environmental monitoring. Moreover, promoting the use of open-source geospatial tools can make technology more accessible, especially for resource-constrained institutions.

Beyond technical training, fostering a culture of data-driven decision-making is crucial. Workshops, seminars, and knowledge-sharing platforms can encourage collaboration and innovation. Furthermore, integrating geospatial education into academic curricula can prepare the next generation of professionals to leverage these technologies effectively.

Investing in capacity building not only enhances the effective use of geospatial data but also ensures sustainability. As technologies evolve, continuous learning and adaptation become essential. By prioritizing training and skill development, India can harness the full potential of geospatial technologies to drive inclusive and sustainable development.

7 International Best Practices

7.1 Singapore's Smart Nation Policy

Singapore's Smart Nation initiative offers a leading example of integrating remote sensing and digital data into planning. Recognizing land scarcity, Singapore built a nationwide 3D city model called Virtual Singapore. Using aerial photography and airborne/terrestrial LiDAR, the government created a detailed 3D digital twin of the city-state. This platform acts as "a dynamic 3D city model and collaborative data platform" enabling planners to visualize interventions safely. For instance, planners in the Yuhua district use Virtual Singapore to evaluate pedestrian bridge designs in context. The platform also ingests real-time sensor feeds: air quality, traffic cameras, and other IoT data from Singapore's extensive sensor network are overlaid on the 3D model. Applications include flood management (civil defence) and infrastructure planning (utilities, transport). The project is driven by high-level policy (National Research Foundation funding) and broad agency cooperation. Singapore's approach shows how a government can leverage satellite/airborne remote sensing plus IoT for comprehensive urban decision-making. Indian cities can draw lessons: a digital twin could similarly enable "test-bed solutions for effective planning" without risking public resources (Prime 2022). The Singapore Land Authority (SLA) has been instrumental in developing geospatial platforms like OneMap and GeoSpace, which provide accessible geospatial data to the public and government agencies, respectively. These platforms facilitate data sharing and collaboration across different sectors, promoting transparency and efficiency in urban governance (Authority 2024). Singapore's approach demonstrates how integrating geospatial technologies into urban planning can lead to more informed decision-making, efficient resource allocation, and enhanced citizen engagement. For countries like India, adopting similar strategies could significantly improve urban governance and infrastructure development.

7.2 Barcelona's Climate-Sensitive Urban Planning

Barcelona has emerged as a leader in climate-sensitive urban planning, integrating environmental considerations into its urban development strategies. The city's approach emphasizes sustainability, resilience, and social equity. One of the key initiatives is the **Superblock program**, which reorganizes city blocks to prioritize pedestrians and cyclists, reduce traffic congestion, and increase green spaces. This initiative not only improves air quality and reduces noise pollution but also fosters community interaction and enhances the overall quality of urban life (Climate-ADAPT 2016). Barcelona's **Climate Plan** aims to reduce greenhouse gas emissions by 45% by 2030 and achieve carbon neutrality by 2050. The plan includes measures such as expanding green infrastructure, promoting sustainable mobility, and enhancing energy efficiency in buildings (Ventayol et al. 2018). The city also utilizes geospatial tools like the **Climate Vulnerability Map**, which identifies areas most susceptible to climate risks. This tool aids in prioritizing interventions and allocating resources effectively to enhance urban resilience (Climate Vulnerability Map 2025). Barcelona's integration of climate considerations into urban planning serves as a model for cities worldwide. For Indian cities facing similar challenges, adopting such climate-sensitive approaches can lead to more sustainable and livable urban environments.

7.3 LIDAR and Planning in North America

In North America, Light Detection and Ranging (LIDAR) technology has become an essential tool in urban planning and infrastructure development. LIDAR provides high-resolution, three-dimensional data that enables precise mapping and analysis of urban environments.

Cities like San Francisco have employed LIDAR to create detailed 3D maps for urban planning purposes. For instance, LIDAR data has been used to assess building heights, analyse terrain, and plan infrastructure projects with greater accuracy (Palmer 2015).

In Toronto, LIDAR has been utilized to model urban vegetation and assess the impact of green spaces on microclimates. This information aids in planning green infrastructure and mitigating urban heat island effects.

Furthermore, LIDAR technology has been instrumental in disaster management. For example, in Miami Beach, LIDAR data has been used to map flood-prone areas and plan mitigation strategies.

The adoption of LIDAR in urban planning across North America highlights its potential to enhance data accuracy, inform decision-making, and improve the resilience of urban infrastructure. For India, integrating LIDAR technology into urban planning processes could significantly advance the precision and effectiveness of development projects.

7.4 Transferable Lessons for India

The international examples of Singapore, Barcelona, and North American cities offer valuable insights for India as it navigates the complexities of urbanization and climate change.

1. **Embracing Geospatial Technologies**: Singapore's integration of geospatial platforms like Virtual Singapore demonstrates the importance of leveraging technology for urban planning. India can adopt similar digital twin models to simulate urban scenarios and optimize infrastructure development (Prime 2022).

2. **Prioritizing Climate Resilience**: Barcelona's climate-sensitive planning underscores the need to incorporate environmental considerations into urban development. Indian cities can implement green infrastructure, promote sustainable mobility, and utilize tools like climate vulnerability maps to enhance resilience (Climate Vulnerability Map 2025).
3. **Utilizing Advanced Mapping Tools**: The use of LIDAR in North America illustrates the benefits of high-resolution mapping for urban planning. India can invest in LIDAR technology to improve the accuracy of spatial data, aiding in infrastructure planning and disaster management.
4. **Fostering Interagency Collaboration**: A common thread in these international practices is the collaboration between various government agencies, academia, and the private sector. India can establish similar partnerships to facilitate data sharing, innovation, and capacity building.

By learning from these global best practices, India can enhance its urban planning processes, promote sustainable development, and build resilient cities equipped to face future challenges.

8 Recommendations for Indian Policy Context

8.1 Integrating Remote Sensing into Master Plans

Integrating remote sensing (RS) and Geographic Information Systems (GIS) into urban master planning is crucial for India's sustainable urban development. RS provides up-to-date satellite imagery, enabling planners to monitor urban sprawl, land-use changes, and environmental impacts effectively.

A study on Arakkonam town in Tamil Nadu demonstrated the efficacy of using RS and GIS for master plan preparation. The study utilized various thematic layers such as soil, geology, and transportation networks to conduct urban suitability analysis, leading to a comprehensive master plan that addressed infrastructure needs based on projected population growth (Shyam Sundar Raj and Venkatesan 2018).

The National Remote Sensing Centre (NRSC) of India has also emphasized the role of RS in urban planning. Their resources highlight how RS can identify urban growth outside formal planning controls, allowing for timely remedial measures to provide necessary infrastructure and improve public health and hygiene (*Remote Sensing Application*).

By adopting RS and GIS technologies, urban planners in India can create dynamic, data-driven master plans that are responsive to real-time changes, thereby enhancing the effectiveness and sustainability of urban development projects.

8.2 *Encouraging PPPs and Innovation Ecosystems*

Public–private partnerships (PPPs) are instrumental in bridging infrastructure gaps and fostering innovation in India's urban development. By leveraging private sector expertise and funding, PPPs can accelerate the implementation of sustainable and technologically advanced urban projects (India's Public–Private Partnerships for Climate are a Global Model to Follow 2024).

The World Economic Forum highlights India's leadership in PPPs for climate-resilient urban development, waste management, and the circular economy. These partnerships are pivotal in achieving India's commitments to net-zero emissions and reducing carbon intensity (Forum 2015).

A study on PPPs for low-carbon infrastructure in India underscores the impact of government policies in encouraging private sector collaboration. The research examines regulations related to waste management, green buildings, and sustainable transportation, assessing their effectiveness in promoting private investment (Bansal et al. 2025).

Furthermore, the Observer Research Foundation discusses the evolution and advantages of the PPP model in India, noting its role in enhancing efficiency and innovation in municipal infrastructure and service delivery (Jha 2024). To maximize the benefits of PPPs, India should continue to refine its policy frameworks, ensuring they are conducive to private sector participation while safeguarding public interests.

8.3 *Creating Urban Geospatial Data Portals*

Developing centralized urban geospatial data portals is essential for informed decision-making and efficient urban governance in India. These platforms facilitate data sharing among various stakeholders, including government agencies, urban planners, and the public.

The Bharat Maps initiative by the National Informatics Centre (NIC) offers a multi-layer GIS platform depicting core foundation data, serving as an integrated base map service for various applications (*BharatMaps*).

Esri India's Urban Observatory on ArcGIS Hub provides GIS-based dashboards that deliver urban analytics, such as drinking water accessibility in Indian cities. These tools enable real-time monitoring and analysis, supporting data-driven urban planning (India Urban Observatory—On ArcGIS Hub 2023). Additionally, the India Urban Data Exchange (IUDX) is an open-source platform facilitating data sharing among smart city stakeholders, promoting the generation of actionable insights through open datasets (NIUA 2023).

By investing in and expanding such geospatial data portals, India can enhance transparency, foster collaboration, and improve the efficiency of urban development initiatives.

8.4 Enhancing Legal Support and Data Governance

Robust legal frameworks and data governance policies are vital for the ethical and secure use of geospatial data in urban planning. India has made significant strides in this area, but continuous improvement is necessary to address emerging challenges.

The Department of Science and Technology (DST) has issued guidelines for acquiring and producing geospatial data, aiming to streamline data sharing and promote innovation (Guidelines for Acquiring and Producing Geospatial Data and Geospatial Data Services Including Maps 2021).

The World Bank's assessment of global data regulation highlights the importance of clear legal frameworks in building trust in the data economy. It emphasizes the need for certainty, rules, and predictability to support data governance (Chen 2021).

Furthermore, the Oxford Academic publication discusses India's data governance landscape, noting the balance between promoting digital innovation and protecting privacy rights. It underscores the necessity for comprehensive policies that address both personal and non-personal data (Mishra 2023).

To enhance data governance, India should focus on developing unified policies that facilitate data interoperability, ensure privacy protection, and promote the ethical use of geospatial information in urban planning.

9 Conclusion

9.1 Summary of Findings

The integration of geospatial data into urban policy and planning in India is crucial for realizing the full potential of data-driven governance (Aayog 2018). Several challenges in the current system, such as fragmentation across departments, legal and regulatory constraints, and the lack of technical capacity within urban local bodies (ULBs), have hindered the effective use of geospatial technologies. Geospatial data, when properly integrated and managed, can significantly enhance urban development initiatives by providing real-time information for decision-making and resource allocation (ISRO 2023). However, overcoming these barriers requires significant reform in policy, data governance, and institutional collaboration.

From the exploration of international best practices, it was evident that countries like Singapore and cities like Barcelona have successfully used geospatial technologies to create more sustainable, resilient, and adaptive urban environments. Singapore's Smart Nation initiative, for example, has leveraged cutting-edge technologies like the Internet of Things (IoT) and Geographic Information Systems (GIS) to enhance urban planning, governance, and service delivery. Similarly, Barcelona's focus on climate-sensitive urban planning demonstrates how geospatial data can be used to address the challenges posed by climate change and urbanization.

India can draw valuable lessons from these examples, particularly in fostering collaborations between government agencies, private sector innovators, and citizens. Additionally, policies that promote open-access geospatial data and encourage the use of Remote Sensing and GIS in urban planning can help overcome data silos and promote transparency and accountability.

9.2 Towards Inclusive and Responsive Urban Futures

The future of urban planning in India must be inclusive, responsive, and adaptable to rapid urbanization and environmental changes. As cities in India continue to expand, there is a growing need for planning systems that can respond dynamically to changes in population, infrastructure, and resource needs. Geospatial data, when integrated with urban policies, can enable a more holistic understanding of urban dynamics and support the development of Smart Cities (Bank 2019).

Inclusive urban futures require an approach that ensures the participation of all stakeholders, including marginalized communities, in the decision-making process. Geospatial technologies can facilitate the involvement of citizens by providing accessible, real-time data and platforms for engagement. For instance, participatory urban planning processes can be enhanced through tools that allow citizens to interact with spatial data and contribute to the planning of their neighbourhoods. By enabling informed decision-making and better resource allocation, these technologies can help ensure that urban development benefits all segments of society.

Furthermore, a responsive urban future depends on creating systems that can quickly adapt to environmental changes and crises, such as floods, earthquakes, and pandemics. With the integration of remote sensing and real-time monitoring, cities can not only improve disaster preparedness and response but also enhance the overall quality of life by ensuring that essential services are provided efficiently and equitably.

9.3 Final Thoughts and Research Opportunities

The integration of geospatial technologies into urban governance in India holds immense potential to improve planning, governance, and service delivery. However, achieving this potential requires overcoming several barriers, including the fragmentation of geospatial data, the lack of coordination among departments, and the need for better data governance frameworks. The implementation of a robust and inclusive geospatial policy will be key to overcoming these challenges and facilitating the integration of geospatial data into urban management processes.

Future research in this area should focus on several key aspects. First, the development of models for inter-agency data sharing and collaboration will be essential in ensuring that geospatial data can be accessed and utilized by all relevant stakeholders.

Second, research on the legal and ethical aspects of geospatial data usage, particularly in relation to privacy concerns and data sovereignty, is necessary to establish clear guidelines and frameworks for the ethical use of such data. Finally, the role of emerging technologies, such as artificial intelligence and blockchain, in enhancing geospatial data management and utilization should be explored.

As India continues its journey towards becoming a global leader in urban innovation, the integration of geospatial data into planning and governance processes will play a pivotal role in shaping the future of its cities. With the right policies, institutional reforms, and technological advancements, India can create urban environments that are not only sustainable and resilient but also equitable and inclusive for all citizens.

References

Aayog N (2018) National strategy for artificial intelligence. https://www.niti.gov.in/sites/default/files/2023-03/National-Strategy-for-Artificial-Intelligence.pdf

Aayog N (2021) Reforms in urban planning capacity in India

Ahluwalia IJ (2019) Urban governance in India. J Urban Aff 41(1):83–102. https://doi.org/10.1080/07352166.2016.1271614

Authority SL (2024) Singapore geospatial master plan 2024–2033. https://file.go.gov.sg/singapore-geospatial-master-plan-2024.pdf

Aziz F, Wang X, Mahmood MQ, Awais M, Trenouth B (2024) Coastal urban flood risk management: challenges and opportunities—a systematic review. J Hydrol 645. https://doi.org/10.1016/j.jhydrol.2024.132271

Bank W (2019) Geospatial technology and information for development. World Bank Group

Bansal D, Bhattacharya N, Ghosal I (2025) Encouraging public private collaborations for low carbon green infrastructure in India: the impact of government policies. In: Hossain IAK, Haque MM, Ullah SMA (eds) Intersecting environmental governance with technological advancements. IGI Global, pp 149–180. https://doi.org/10.4018/979-8-3693-7001-8.ch006

Chen R (2021) Mapping data governance legal frameworks around the world. The World Bank

Cheval S, Amihăesei VA, Chitu Z, Dumitrescu A, Falcescu V, Irașoc A, Micu DM, Mihulet E, Ontel I, Paraschiv MG, Tudose NC (2024) A systematic review of urban heat island and heat waves research (1991–2022). Clim Risk Manag 44. https://doi.org/10.1016/j.crm.2024.100603

Climate-ADAPT (2016) Barcelona: trees tempering the Mediterranean city climate. https://climate-adapt.eea.europa.eu/en/metadata/case-studies/barcelona-trees-tempering-the-mediterranean-city-climate

Climate Vulnerability Map (2025) https://chatgpt.com/c/6818fab1-28dc-8000-b4b2-07bf77f7357b

Communications GM (2023) National geospatial policy aims to transform Indian innovation by 2035. Geospatial World. https://geospatialworld.net/prime/business-and-industry-trends/national-geospatial-policy-indian-innovation-2035/?utm_source=chatgpt.com

Contributors W (2025) ISRO facilities. https://en.wikipedia.org/wiki/ISRO_facilities

Forum WE (2015) How is open data changing India? https://www.weforum.org/stories/2015/02/how-is-open-data-changing-india/

Group WB (2021) Geospatial technology and information for development. https://www.worldbank.org/en/topic/land/brief/geospatial-technology-and-information-for-development

Guidelines for Acquiring and Producing Geospatial Data and Geospatial Data Services including Maps (2021)

India MoUDGo (2015) Mission statement and guidelines. https://scbp.niua.org/sites/default/files/AMRUT%20Guidelines%202015.pdf
India Urban Observatory—On ArcGIS Hub (2023) EsRI India. https://www.esri.in/en-in/newsroom/blog/india-urban-observatory-on-arcgis-hub
India's Public-Private Partnerships for Climate are a Global Model to Follow (2024) World Economic Forum. https://www.weforum.org/stories/2024/01/india-public-private-partnerships-climate/
ISRO (2023) Bhuvan geoportal overview. National Remote Sensing Centre
Jha R (2024) Improving PPP strategies for municipal infrastructure and service delivery, p 724. O. R. Foundation. https://www.orfonline.org/public/uploads/posts/pdf/20240809094003.pdf
Li F, Yigitcanlar T, Nepal M, Nguyen K, Dur F (2023) Machine learning and remote sensing integration for leveraging urban sustainability: a review and framework. Sustain Cities Soc 96. https://doi.org/10.1016/j.scs.2023.104653
Marasinghe R, Yigitcanlar T, Mayere S, Washington T, Limb M (2024) Computer vision applications for urban planning: a systematic review of opportunities and constraints. Sustain Cities Soc 100. https://doi.org/10.1016/j.scs.2023.105047
Mishra N (2023) Data governance and digital trade in India: losing sight of the forest for the trees? In: Chander A, Sun H (eds) Data sovereignty: from the digital silk road to the return of the state. Oxford University Press. https://doi.org/10.1093/oso/9780197582794.003.0011
National Data Sharing and Accessibility Policy (2012) New Delhi. https://dst.gov.in/sites/default/files/gazetteNotificationNDSAP.pdf
National Geospatial Policy (2022) https://pib.gov.in/PressReleasePage.aspx?PRID=2106569
NIUA (2023) How India leverages geospatial technologies for urban management. CDG. https://niua.in/cdg/news-listing/how-india-leverages-geospatial-technologies-urban-management
Palmer KM (2015) Bombarding San Francisco with lasers to create a perfect 3-D map. WIRED. https://www.wired.com/2015/11/bombarding-san-francisco-with-lasers-to-create-a-perfect-3-d-map
Perveen S, Kamruzzaman M, Yigitcanlar T (2017) Developing policy scenarios for sustainable urban growth management: a Delphi approach. Sustainability 9(10):1787
Prime G (2022) Virtual Singapore—building a 3D-empowered smart nation. Geospatial World. https://geospatialworld.net/prime/case-study/national-mapping/virtual-singapore-building-a-3d-empowered-smart-nation
Remote Sensing Application, National Remote Sensing Centre. https://www.nrsc.gov.in/sites/default/files/pdf/ebooks/Chap_5_Urban.pdf
Shyam Sundar Raj RMR, Venkatesan C (2018) Master plan preparation for Arakkonam town using remote sensing and GIS. J Civ Eng Constr Technol 7(3):115. https://core.ac.uk/download/230812736.pdf
Son TH, Weedon Z, Yigitcanlar T, Sanchez T, Corchado JM, Mehmood R (2023) Algorithmic urban planning for smart and sustainable development: systematic review of the literature. Sustain Cities Soc 94. https://doi.org/10.1016/j.scs.2023.104562
Suprijanto A, Tan Y, Lodhi MK, Santillan RDM, Alage IL, Zamalloa GRP (2025) Industrial heat island and carbon emissions: machine learning-based environmental impact assessment in Cilegon, Indonesia. Environ Chall 19. https://doi.org/10.1016/j.envc.2025.101141
Technology DoS (2021a) National geospatial programme (erstwhile NRDMS). https://dst.gov.in/national-geospatial-programme-erstwhile-nrdms
Technology DoS (2021b) National spatial data infrastructure (NSDI) initiative: a decade of geospatial data sharing and integration in India. https://www.nsdiclearinghouse.gov.in/erdasapollo/nsdiportal/meetings/NSDI_Corrected_22nd%20Oct.pdf
Veibiakkim R, Shkaruba A, Sepp K (2025) A systematic review of urban ecosystem disservices and its evaluation: key findings and implications. Environ Sustain Indic 26. https://doi.org/10.1016/j.indic.2025.100612
Ventayol I, Castells C, Font Ferrer T, Franquesa T, Gabàs A, Soldevilla I (2018) Climate plan 2018–2030. https://www.barcelona.cat/barcelona-pel-clima/sites/default/files/climate_plan_maig.pdf

Wellmann T, Lausch A, Andersson E, Knapp S, Cortinovis C, Jache J, Scheuer S, Kremer P, Mascarenhas A, Kraemer R, Haase A, Schug F, Haase D (2020) Remote sensing in urban planning: contributions towards ecologically sound policies? Landsc Urban Plan 204:103921. https://doi.org/10.1016/j.landurbplan.2020.103921

Wulder MA, Loveland TR, Roy DP, Crawford CJ, Masek JG, Woodcock CE, Allen RG, Anderson MC, Belward AS, Cohen WB, Dwyer J, Erb A, Gao F, Griffiths P, Helder D, Hermosilla T, Hipple JD, Hostert P, Hughes MJ, Zhu Z (2019) Current status of Landsat program, science, and applications. Remote Sens Environ 225:127–147. https://doi.org/10.1016/j.rse.2019.02.015

Index

B. Kozhikkodan Veettil and B. Das (eds.), *Remote Sensing of Urban Heat Islands*, Cities and Nature, https://doi.org/10.1007/978-3-032-11580-5

MIX
Papier aus verantwortungsvollen Quellen
Paper from responsible sources
FSC® C105338

If you have any concerns about our products,
you can contact us on
ProductSafety@springernature.com

In case Publisher is established outside the EU,
the EU authorized representative is:
Springer Nature Customer Service Center GmbH
Europaplatz 3, 69115 Heidelberg, Germany

Printed by Libri Plureos GmbH
in Hamburg, Germany